Overcoming Barriers for Women of Color in STEM Fields:

Emerging Research and Opportunities

Pamela M. Leggett-Robinson
PLR Consulting, USA

Brandi Campbell Villa
Belay Consulting, USA

A volume in the Advances in Human and Social Aspects of Technology (AHSAT) Book Series

IGI Global
PUBLISHER of TIMELY KNOWLEDGE

Published in the United States of America by
IGI Global
Information Science Reference (an imprint of IGI Global)
701 E. Chocolate Avenue
Hershey PA, USA 17033
Tel: 717-533-8845
Fax: 717-533-8661
E-mail: cust@igi-global.com
Web site: http://www.igi-global.com

Library of Congress Cataloging-in-Publication Data

Names: Leggett-Robinson, Pamela M., 1968- editor.
Title: Overcoming barriers for women of color in STEM fields : emerging research and opportunities / Pamela M. Leggett-Robinson and Brandi Villa, editors.
Description: Hershey, PA : Information Science Reference, [2021] | Includes bibliographical references and index. | Summary: "This book addresses the formidable barriers faced and overcome by women of color in STEM, as well as how the navigation of the STEM landscape impacts their lives"-- Provided by publisher.
Identifiers: LCCN 2020002722 (print) | LCCN 2020002723 (ebook) | ISBN 9781799848585 (hardcover) | ISBN 9781799853763 (paperback) | ISBN 9781799848592 (ebook)
Subjects: LCSH: Women in science. | African American women in the professions. | Women scientists. | Feminism and science. | Professions--Women.
Classification: LCC Q130 .O89 2021 (print) | LCC Q130 (ebook) | DDC 305.43/5--dc23
LC record available at https://lccn.loc.gov/2020002722
LC ebook record available at https://lccn.loc.gov/2020002723

This book is published in the IGI Global book series Advances in Human and Social Aspects of Technology (AHSAT) (ISSN: 2328-1316; eISSN: 2328-1324)

British Cataloguing in Publication Data
A Cataloguing in Publication record for this book is available from the British Library.

All work contributed to this book is new, previously-unpublished material.
The views expressed in this book are those of the authors, but not necessarily of the publisher.

For electronic access to this publication, please contact: eresources@igi-global.com.

Advances in Human and Social Aspects of Technology (AHSAT) Book Series

ISSN:2328-1316
EISSN:2328-1324

Editor-in-Chief: Ashish Dwivedi, The University of Hull, UK

MISSION

In recent years, the societal impact of technology has been noted as we become increasingly more connected and are presented with more digital tools and devices. With the popularity of digital devices such as cell phones and tablets, it is crucial to consider the implications of our digital dependence and the presence of technology in our everyday lives.

The **Advances in Human and Social Aspects of Technology (AHSAT) Book Series** seeks to explore the ways in which society and human beings have been affected by technology and how the technological revolution has changed the way we conduct our lives as well as our behavior. The AHSAT book series aims to publish the most cutting-edge research on human behavior and interaction with technology and the ways in which the digital age is changing society.

COVERAGE

- Technology Adoption
- ICTs and human empowerment
- Technology and Freedom of Speech
- Cultural Influence of ICTs
- Computer-Mediated Communication
- Gender and Technology
- Technology and Social Change
- Technology Dependence
- Technoself
- Human-Computer Interaction

IGI Global is currently accepting manuscripts for publication within this series. To submit a proposal for a volume in this series, please contact our Acquisition Editors at Acquisitions@igi-global.com or visit: http://www.igi-global.com/publish/.

The Advances in Human and Social Aspects of Technology (AHSAT) Book Series (ISSN 2328-1316) is published by IGI Global, 701 E. Chocolate Avenue, Hershey, PA 17033-1240, USA, www.igi-global.com. This series is composed of titles available for purchase individually; each title is edited to be contextually exclusive from any other title within the series. For pricing and ordering information please visit http://www.igi-global.com/book-series/advances-human-social-aspects-technology/37145. Postmaster: Send all address changes to above address.

Titles in this Series

For a list of additional titles in this series, please visit:
http://www.igi-global.com/book-series/advances-human-social-aspects-technology/37145

Maintaining Social Well-Being and Meaningful Work in a Highly Automated Job Market
Shalin Hai-Jew (Kansas State University, USA)
Business Science Reference • © 2020 • 333pp • H/C (ISBN: 9781799825098) • US $215.00

ICTs and Innovation for Didactics of Social Sciences
Emilio José Delgado-Algarra (University of Huelva, Spain)
Information Science Reference • © 2020 • 295pp • H/C (ISBN: 9781799828822) • US $175.00

Civic Engagement Frameworks and Strategic Leadership Practices for Organization Development
Susheel Chhabra (Periyar Management and Computer College, India) and Muneesh Kumar (Maharaja Agrasen Institute of Technology (MAIT), Delhi, India)
Information Science Reference • © 2020 • 314pp • H/C (ISBN: 9781799823728) • US $195.00

Civic Engagement in Social and Political Constructs
Susheel Chhabra (Periyar Management and Computer College, India)
Information Science Reference • © 2020 • 285pp • H/C (ISBN: 9781799823643) • US $195.00

Recent Advances in Digital Media Impacts on Identity, Sexuality, and Relationships
Michelle F. Wright (Penn State University, USA)
Information Science Reference • © 2020 • 347pp • H/C (ISBN: 9781799810636) • US $195.00

Human Capital Formation for the Fourth Industrial Revolution
Sulaiman Olusegun Atiku (Namibia University of Science and Technology, Namibia)
Business Science Reference • © 2020 • 385pp • H/C (ISBN: 9781522598107) • US $215.00

IGI Global
PUBLISHER of TIMELY KNOWLEDGE

701 East Chocolate Avenue, Hershey, PA 17033, USA
Tel: 717-533-8845 x100 • Fax: 717-533-8661
E-Mail: cust@igi-global.com • www.igi-global.com

Editorial Advisory Board

Table of Contents

Foreword

I was absolutely devastated and humiliated. It was March 2014 and my department chair walked into my office and handed me an envelope, which essentially sealed my fate as an assistant chemistry professor at a small liberal arts institution in the Midwest. After I opened the envelope, I quickly glanced over the document, which indicated that I was denied tenure.

That particular spring day is essentially a blur now, but I do remember texting my personal BOD (Board of Directors) about the tenure denial decision. My BOD is a network of amazing women in STEM, who hold me accountable and tell me the truth, when I need to hear it. I also emailed my colleagues outside of my department, who had been very supportive of me. I was actually in shock about the decision, but thankful that I had a two-week spring break to try to get my emotions under control and figure out just what the hell I was going to do next in my career. I had no intentions of appealing the decision either. For what? To be denied again? Tenure denial was an enormous barrier for me to personally overcome, but I did it. How did I navigate the STEM landscape? I was able to move forward because I am surrounded by a network of amazing women (and men) in both industry and academia who encouraged me to keep going after this awful career setback.

The women celebrated in *Overcoming Barriers for Women of Color in STEM: Emerging Research and Opportunities* have faced tremendous challenges in STEM—but they persisted! Powerful narratives in *Overcoming Barriers* explore diverse topics including the significance of HBCUs (Historically Black Colleges and Universities) and the relationship between racial identity, science identity, and science self-efficacy of African American students attending these institutions; the importance of networking, collaborating and scholarship in research; effective mentoring in STEM; examination of diversity and inclusion initiatives and the STEM professoriate and discovering your own strength and resilience along your STEM journey.

More importantly, this book is a valuable resource for women of color to navigate their own STEM journeys. I absolutely applaud these women for sharing their STEM testimony (Storytelling), which will provide tools and tips for other women facing their own challenges. Overcoming career barriers is certainly a difficult process for anyone. However, the women celebrated in this important book provide a "STEM career road map" to help you walk in your purpose.

Sibrina Collins
Marburger STEM Center, Lawrence Technological University, USA

Preface

Under-representation of women of color (WOC) in STEM is not a new area of research. Relational demography, organizational justice, social network, and critical race theories all identify possible precursors that are deemed responsible for the shortage of women of color in STEM fields. According to the National Science Foundation, underrepresented women refer to Hispanics, Native Americans, and Blacks. These groups have a smaller representation in STEM fields than in the U.S. population.

Over the past 20 plus years, government agencies, national program initiatives, higher education organizations, and non-profit organizations have worked tirelessly to recruit and retain girls (and women) of color in science, technology, engineering, and mathematics (STEM) disciplines. These agencies, initiatives, and organizations (nationwide and online) have led the charge in programming that specifically focuses on getting girls interested and engaged in STEM. Programs run the gamut of afterschool STEM activities, STEM Saturday school, STEM summer camps, and STEM pre-college camps and research opportunities. Many have also narrowed their programming to focus on the different aspects of the development of women as scientists as well as the different disciplines themselves, i.e. Girls who Code, Latinas In STEM, Black Girls Code, STEM Gems, Girls Inc. “Operation SMART”, etc. Some programs aim to make the work climate more friendly to women, while others build strong networks for women through role models and mentoring.

Although well-intentioned, many of these STEM initiatives focusing on girls and women of color ignore the complexity of systemic and interlocking barriers in place and equate the experiences of White girls and women with all girls and women in STEM, thus creating a band-aid approach. These approaches neglect the individual and combined impact of variables such as institutional racism, cultural STEM racism, and gendered racism on educational experiences and outcomes of historically marginalized groups. Given the history and magnitude of race and gender inequities in STEM, it stands to reason that no one program or organization, even national in scope, can solve the problem of marginalization in STEM and address the broader societal issues that underlie it. A close examination of issues faced by girls and women of color and a willingness to make substantive, permanent, systematic changes are steps towards addressing critical barriers to minority participation in STEM.

STEM INEQUITY IS A SOCIAL JUSTICE ISSUE

Disparities in STEM cut across racial, gender, and socioeconomic lines impacting predominantly historically marginalized groups. It is a consistent narrative that the vast majority of students being filtered out of STEM (K-12, post-secondary, graduate, and workplace) are Black and Brown females. In recent years, STEM has been linked to vocational and educational opportunities, and more importantly to earning potential over one's lifetime. Those who are fortunate enough to traverse the STEM landscape are therefore afforded the benefits of attaining better, more satisfying, and higher paying jobs. In sum, filtering Black and Brown female students out of STEM can profoundly impact the trajectory of their lives.

Educational and workplace STEM settings serve as arenas where social inequities and injustices can be produced and reproduced by privileging some social identities while marginalizing others (Roland, 2018; Collins & Bilge, 2016). Numerous factors, including gendered stereotypes, institutional culture, and chilly climates, conspire against and undermine many young Black women's abilities to develop and maintain an interest in STEM and develop their identities in STEM (Brickhouse et al., 2000; Charleston, Adserias, Lang, & Jackson, 2014).

Broadening participation for women of color in STEM is social equity work; intersectionality is at the heart. Intersectionality, as conceptualized by Kimberlé Crenshaw (1989, 1991), is an analytical tool for exposing interlocking structural systems of dominance and subordination such as racism, classism, sexism, and ableism. It offers a way of thinking about social identities and their relationships to power in organizations and institutions.

Because the intersectional experience is greater than the sum of racism and sexism, any analysis that does not take intersectionality into account cannot sufficiently address the particular manner in which Black women are subordinated. (Crenshaw, 1989)

Intersectionality is valuable as a framework because it is not meant to be solely theoretical; it is a critique that fosters conversations for real-world change and progress. (Mitchell, 2014)

WOMEN OF COLOR AND INTERSECTIONALITY

Women tend to be viewed as a homogeneous group of individuals with similar needs, experiences, and barriers to academic and career progress. As such, many studies have treated gender and race separately, obscuring their complex intersection in the context of racial minorities' unique educational, professional, and societal histories.

In STEM, race beyond the discussion of Black and Brown men and gender beyond the topic of White women reveals an often neglected, yet individualized perspective, which exposes subtle, or at times blatant, institutionalized barriers for WOC in STEM (Hall, 2004; Farinde & Lewis, 2012).

Few researchers have examined how the dual presence of race and gender affect the educational experiences of these marginalized women. Seldom are these two features and their simultaneous effects thoroughly explored so that certain truths may be revealed about the Black female's STEM experience as it relates to the salience of her identity and professional success. Although the development of strong, intersectional identities have been identified as critical cultural and societal factors in STEM development (Rosenthal et al., 2011), the intersections of Black women's racial, gender, and scientific identities may conflict with many of the messages they receive throughout their STEM journey and may thus pose a significant challenge to their ability to successfully develop a Black woman scientist identity.

Through a lens of intersectionality, researchers discover not only how Black female students' identities are internally formed and understood, but also how their identities inform their social interactions in STEM. Further exploration can be utilized to address and solve for barriers which promote a lack of STEM identity and belonging for Black girls and women and provide strategies for institutional and organizational STEM social action through theory and practice (Case, 2016; Roland, 2018).

LEAKY PIPES

The STEM pipeline is a metaphor used by the STEM community to describe the linear progression of women and minorities from K-12 to professional careers in STEM. The pipeline metaphor commonly uses the terms "leaks" to describe retention problems which are due to structural, societal, and educational inequities within the STEM culture.

These barriers exist at every stage of STEM progression for women of color and create opportunities to further marginalize their existence in STEM, or cause them to leave STEM altogether.

- **Transition From K-12 to Higher Education Systems**. Published literature emphasizes preparedness deficiencies at both early and secondary levels of education and lack of early STEM career exposure as integral barriers to sustaining Black women STEM scholars in higher levels of academia (Ehrenberg, 2010; George, Neale, Van Horne, & Malcolm, 2001).
- **Post-Secondary Higher Education Systems**. Studies point to social factors of isolation, academic difficulties, and STEM belonging as the primary barriers for departure from STEM (Buzzetto-More et al., 2010; Charleston, 2012; George et al., 2001).
- **Graduate, Postdoctoral, or Junior Level Career**. WOC at this phase of progression are often faced with cultural boundaries that discourage their ability to amalgamate their other, and often-conflicting gender, racial, and academic identities (Syed & Chemers, 2011). Additionally, WOC face the unique problem of balancing their career advancement and their family lives while upholding culturally acceptable roles for their gender as well as their race (Cech, Rubineau, Silbey, & Seron, 2011; Ko et al., 2013).
- **Senior Level Career**. During the senior level, WOC typically develop a strong desire for activism. Ko, Kachchaf, Ong, and Hodari (2013) found that many WOC in senior levels of STEM organizations express a strong desire to improve conditions for younger generations of marginalized women through recruitment, scholarly works drawing attention to the STEM culture, volunteerism, or STEM storytelling. Many of these activities are met with blatant opposition, micro-aggressive statements, promotion denial, or termination. The lack of voice, or the sense that one does not have agency and influence within an organizational context or in this case, a STEM context is yet another barrier. This barrier leads to professional reflection often rendering exodus from the STEM organization as well as a traditional STEM career.

Women of color progress through the STEM pipeline of persistent barriers and inequitable treatment with the assistance of other women who are willing to address hostile STEM educational and workplace environments and share personal experiences of their journey. Both tactics are necessary to slow down or stop the "leaks" of the pipeline and empower women of color to remain in STEM.

EMPOWERMENT THROUGH STORYTELLING

Storytelling is one of the many cultural expressions shared by people throughout history. Before there was even written language, community stories, customs, and beliefs were passed down from generation to generation through stories. More specifically, Black and Brown communities used storytelling as a way to record experiences and to teach the younger generation about their collective cultural past (Santos, 2018). Storytelling was imperative in Black and Brown communities as many were unable to record their experiences due to illiteracy. Throughout history, these communities told stories to record events, cope with insufferable, unjust experiences, and give hope for a better life.

Today, stories (written or oral) continue to give historical reference, teach life lessons, motivate, and serve as life guides. While the metaphors and anecdotes in stories help relate ideas to shared experiences, storytelling allows us to be present with our experience, and to draw others into that experience as well (Coleman, 2013). Just as in the Black and Brown communities of the past, storytelling by women of color in STEM today brings meaning to their experiences.

Storytelling in reference to one's own experiences is considered a personal narrative (or disclosure) and allows one to sort through events, thoughts, and feelings that have caused trauma. For women of color in STEM, trauma takes the form of isolation, microaggressions, gender and racial bias, tokenism, and racial/ethnic stereotyping.

Disclosure (written and oral) of such trauma has been associated with improvements to one's mental and physical health, i.e., decreased distress, depression, and fewer illness-related visits to the physicians (Donnelly, 1991; Ullrich, 2002). Additionally, expressing trauma-related emotions through storytelling (written or oral disclosure), enhances feelings of control and mastery over the trauma, leads to improved self-reported health, psychological well being, physiological functioning, and can be used as an ongoing coping strategy for managing emotional and psychosocial turmoil (Smith, 1998; Holloway & Freshwater, 2007).

For women of color in STEM, storytelling (disclosure) is a powerful tool. Along with the aforementioned, it allows marginalized groups to reflect and verbalize their own trauma while simultaneously revealing their authenticity and vulnerability. For younger women hearing the stories, it diminishes feelings of isolation and fear of telling their own stories. Additionally, it is through the lens of "story" that women of color begin to face STEM culture trauma, embrace their vulnerabilities, and empower themselves.

Owning our story can be hard but not nearly as difficult as spending our lives running from it. Embracing our vulnerabilities is risky but not nearly as dangerous as giving up on love and belonging and joy — the experiences that make us the most vulnerable. Only when we are brave enough to explore the darkness will we discover the infinite power of our light. (Brown, 2012)

To truly address the institutional, educational, cultural, and societal barriers that continue to limit representation of WOC in STEM, their stories must be told as non-commercialized narratives. Their narrative should be able to expose the reluctance, exclusion, mis-opportunities, and under-education of their journey in STEM. More importantly, these narratives should shed light on the institutions, organizations, and industries responsible for perpetuating and participating in STEM inequity.

When WOC hear and process stories (narratives) by other women of color, they hear stories about adjusting to disabling or hostile working conditions, living with memories of discrimination and micro-aggressions, caregiving challenges, and professional changes. Most importantly they hear about learning how to cope with what hand life has dealt them and reflections of hope and transformation throughout the process. Through authentic sharing of stories, women of color in STEM are able to be vulnerable, be seen, and be heard. Everyone has a story to tell, and everyone deserves a chance to tell their story from their perspective - in this case, being a Black or Brown woman in STEM.

OUR PURPOSE

In the face of institutional and societal structural bias, it is important to understand the tools women of color use to navigate the STEM landscape. Alternatively, it is equally as important to understand the role of intersectionality in their personal and professional identities in overcoming the systemic (intentional or unintentional) barriers placed before them. As most published resources regarding WOC in STEM focus solely on overcoming the barriers in STEM, this book exposes the introspection of WOC (professional and personal scars and celebrations) obtained by overcoming STEM barriers and the impact left on the STEM community at large through research studies and personal narratives (storytelling). The inspirational transformation within their personal and professional lives spans across the spectrum of all women of color in STEM. More importantly, it gives a perspective of understanding to continually suppressed and rarely expressed fears that limit empowerment of WOC and keep them invisible within the contexts of STEM.

The contributions from the authors are relevant because, despite a plethora of initiatives, policies, and procedures to increase their representation in STEM, women of color remain underrepresented – their stories are overlooked, their voices silenced, and their transformations while navigating the landscape are rarely acknowledged or celebrated. Through a reflective lens, the narratives of these difficulties will be unpacked, exposed, and examined. The narratives are akin to the Butterfly effect, a metamorphosis that brings about a dramatic change in character and perspective for those who go through it. The Butterfly effect for WOC in STEM is about reflection, rebirth, evolution, and renewal. It sheds light on the courage to confront fear, take leaps, make change, and celebrate their process.

Each chapter exposes the reader to barriers that contribute to the under-representation of WOC in STEM fields but also gives insight to finding meaning in overcoming the barriers. The contributors give factual evidence and personal perspective of discriminatory practices and bias within the culture of STEM. The contributors offer practical strategies to combat feelings of isolation, invisibility, and fear of belonging in STEM at the intersection of gender and race, while also making meaning from the benefits of struggle.

Used as a resource in African American and Gender studies courses, professional and personal libraries, STEM book clubs, mentoring organizations, this text will provide a jump start to courageous conversations and discussions around being a woman (or girl) of color in STEM without fear. This text is a must have for everyone (academicians, researchers, K-12 teachers, mentors, thought leaders, industry leaders, and professional organizations) seeking to empower, encourage, and give voice to women and girls of color in STEM.

ORGANIZATION OF THE BOOK

The book is organized into seven chapters. A brief description of each of the chapters follows:

Chapter 1 introduces the concepts of critical race theory, permanence of racism, and intersectionality. The chapter sets the tone for the contributions by the various authors. In particular, this chapter speaks to the larger issue of how each impacts the racial and science identities of Black women in STEM at both the secondary and postsecondary educational levels. The chapter examines existing statistical data on women of color in STEM, offers definitions for critical race theory (CRT) tenets and constructs within the multidimensional inventory of Black identity (MIBI) to address the importance of racial identity. Additionally, the chapter gives a comparison of institutional contexts (HBCUs vs. PWIs) and their role in developing identities of Black women in STEM.

Chapter 2 examines the impact of segregation at the K-12 level on STEM education and the children in the NYC school system. The chapter gives a historical overview of the effects of *Brown v Education* on the New York City School system (in totality) and takes a comprehensive look at the impact segregation (educational outcomes) has had on African American girls within the NYC school system. This chapter reports STEM achievement gaps for African American girls in K-12. Recommendations addressing the achievement gaps for African American girls are shared.

Chapter 3 employs a STEM autobiography to highlight the familiar and formidable barriers faced by women of color in STEM fields. The chapter reminds readers of the often overlooked or buried feelings of isolation, academic or professional unworthiness, and race/gender bias found in early STEM encounters. The chapter offers the counter space of storytelling to process feelings of STEM related trauma, bring meaning to STEM experiences, and celebrate the professional and personal transformations of navigating the STEM landscape. The author compares the celebratory transformation of women of color in STEM to that of a butterfly - moving from one dimension to another and undergoing a deep experience that leads to an entirely new self-capacity and perception.

Chapter 4 explores the importance of networking for women of color in STEM as it pertains to career advancement. The chapter provides examples of various ways women of color can network to create opportunities that are more than simply ‘leaning in’ or ‘being more aggressive.’ In recounting her experiences, the author discusses the importance of networking, effective networking, networking buckets, and building one’s own network. Strategies for successful networking are provided.

Chapter 5 discusses definitions of diversity and institutional reporting of diversity gains as well as addresses the struggles of post-secondary institutions with identifying the appropriate strategies and/or interventions necessary to actively recruit and retain women faculty of color in STEM. The chapter highlights the barriers found in STEM departments within the academy (microaggressions, racial and gender bias, hostile classrooms, and devalued scholarship), the navigation for successful tenure and promotion, and the personal cost of obtaining tenure and promotion. The chapter uses an autoethnographic approach to reflect on the experience of a Black woman in STEM working at both a historical Black university (HBCU) and a liberal arts intensive predominantly White institution. Recommendations for the retention of minority faculty at PWIs are shared.

Chapter 6 details the mentoring experiences of a Black woman throughout her STEM journey. A literature review of stereotypical images of scientists and their impact on the knowledge, belief, and expectations of black females in STEM is discussed. The chapter highlights the importance of mentors, role models, the HBCUs (community) and systems of support in preparing Black women for STEM careers. Additionally, the chapter discusses how role models and mentors disrupt the narrative of “less-than” for Black females in STEM and ways in which they provide access to wisdom and survival skills. In recounting her STEM experiences, the author reflects on the role of mentoring played in her life and how she pays it forward.

Chapter 7 shares the lived experience and reflection of being a Black woman in an academic engineering department. The chapter incorporates “sanitized” excerpts from the author’s journals, providing real-life thoughts during a 30 year STEM journey. The chapter brings to light on the importance of personal journaling for women of color in STEM as a means to rediscover one's original self, reset the roadmap, revitalize the game plan, regain confidence, and restore resilience. The overarching goal of the chapter is to compel the reader to examine their own journey and identify, create, and execute a revitalized plan to fully experience each moment of success while putting the challenging experiences in the proper perspective. Both STEM Success Cycle and STEM Career box activities are provided to jump start personal journaling.

Pamela M. Leggett-Robinson
PLR Consulting, USA

REFERENCES

Brickhouse, N. W., Lowery, P., & Schultz, K. (2000). What kind of a girl does science? The construction of school science identities. *Journal of Research in Science Teaching*, *37*(5), 441–458. doi:10.1002/(SICI)1098-2736(200005)37:5<441::AID-TEA4>3.0.CO;2-3

Brown, B. (2012). *Daring greatly: how the courage to be vulnerable transforms the way we live, love, parent, and lead*. New York, NY: Gotham Books.

Buzzetto-More, N., Ukoha, O., & Rustagi, N. (2010). Unlocking the barriers to women and minorities in computer science and information systems studies: Results from a multimethodolical study conducted at two minority serving institutions. *Journal of Information Technology Education*, *9*, 115–131. doi:10.28945/1167

Case, K. (Ed.). (2016). *Intersectional pedagogy: Complicating identity and social justice*. Routledge. doi:10.4324/9781315672793

Cech, E., Rubineau, B., Silbey, S., & Seron, C. (2011). Professional role confidence and gendered persistence in engineering. *American Sociological Review*, *76*(5), 641–666. doi:10.1177/0003122411420815

Charleston, L. J. (2012). A qualitative investigation of African Americans' decision to pursue computing science degrees: Implications for cultivating career choice and aspiration. *Journal of Diversity in Higher Education*, *5*(4), 222–243. doi:10.1037/a0028918

Charleston, L. J., Adserias, R. P., Lang, N. M., & Jackson, J. F. L. (2014). Intersectionality and STEM: The role of race and gender in the academic pursuits of African American women in STEM. *Journal of Progressive Policy and Practice*, *2*(3), 273–293.

Coleman, F. (2013, June 25). Only Connect: Why Your Story Matters. *HuffPost*. https://www.huffpost.com/entry/only-connect-why-your-story-matters_b_3492718

Collins, P., & Bilge, S. (2016). *Intersectionality*. John Wiley & Sons.

Crenshaw, K. (1989). Demarginalizing the intersection of race and sex: A Black feminist critique of antidiscrimination doctrine, feminist theory, and antiracist politics. *University of Chicago Legal Forum, 139*, 139–167.

Crenshaw, K. (1991). Mapping the margins: Intersectionality, identity politics, and violence against women of color. *Stanford Law Review, 43*(6), 1241–1299. doi:10.2307/1229039

Donnelly, D. A., & Murray, E. J. (1991). Cognitive and emotional changes in written essays and therapy interventions. *Journal of Social and Clinical Psychology, 10*(3), 334–350. doi:10.1521/jscp.1991.10.3.334

Ehrenberg, R. G. (2010). Analyzing the factors that influence persistence rates in STEM field, majors: Introduction to the symposium. *Economics of Education Review, 29*(6), 888–891. doi:10.1016/j.econedurev.2010.06.012

Farinde, A., & Lewis, C. (2012). The Underrepresentation of African American Female Students in STEM Fields: Implications for Classroom Teachers. *US-China Education Review B, 4*, 421–430.

George, Y. S., Neale, D. S., Van Horne, V., & Malcolm, S. M. (2001). *In pursuit of a diverse science, technology, engineering, and mathematics workforce*. Washington, DC: American Association for the Advancement of Science. Retrieved from http://ehrweb.aaas.org/mge/Reports/Report1/AGEP/?downloadURL=true&loId=EB79A2C2-3280-4404-AAF3-0D5D3F8A9D6D

Hall, R. E. (2004). Entitlement disorder: The colonial traditions of power as white males resistance to affirmative action. *Journal of Black Studies, 34*(4), 562–579. doi:10.1177/0021934703259013

Holloway, I., & Freshwater, D. (2007). Vulnerable story telling: Narrative research in nursing. *Journal of Research in Nursing, 12*(6), 703–711. doi:10.1177/1744987107084669

Ko, L. T., Kachchaf, R. R., Ong, M., & Hodari, A. K. (2013). Narratives of the double bind: Intersectionality in life stories of women of color in physics, astrophysics and astronomy. In P. V. Engelhardt, A. D. Churukian, & N. S. Rebello (Eds.), *Proceedings of the 2012 Physics Education Research Conference* (Vol. 1513, pp. 222–225). Melville, NY: AIP. 10.1063/1.4789692

Mitchell, D. Jr. (2014). Introduction. In D. Mitchell Jr, C. Simmons, & L. Greyerbiehl (Eds.), *Intersectionality & higher education: Theory, research, and praxis* (pp. 1–5). New York, NY: Peter Lang. doi:10.3726/978-1-4539-1407-6

Roland, E. (2018). Understanding Intersectionality to Promote Social Justice in Educational Leadership: Review of JCEL Cases. *Intersections: Critical Issues in Education,* 2, 1. https://digitalrepository.unm.edu/intersections/vol2/iss1/3

Rosenthal, L., London, B., Levy, S. R., & Lobe, M. (2011). The Roles of Perceived Identity Compatibility and Social Support for Women in a Single-Sex STEM Program at a Co-educational University. *Sex Roles*, *65*(9-10), 725–736. doi:10.100711199-011-9945-0

Santos, J. (2018, October 8). How African American folklore saved the cultural memory and history of slaves. *The Conversation*. https://theconversation.com/how-african-american-folklore-saved-the-cultural-memory-and-history-of-slaves-98427

Smyth, J. M. (1998). Written emotional expression: Effects sizes, outcome types, and moderating variables. *Journal of Consulting and Clinical Psychology*, *66*(1), 174–184. doi:10.1037/0022-006X.66.1.174 PMID:9489272

Syed, M., & Chemers, M. M. (2011). Ethnic minorities and women in STEM: Casting a wide net to address a persistent social problem. *The Journal of Social Issues*, *67*(3), 435–441. doi:10.1111/j.1540-4560.2011.01708.x

Ullrich, P. M., & Lutgendorf, S. K. (2002). Journaling about stressful events: Effects of cognitive processing and emotional expression. *Annals of Behavioral Medicine*, *24*(3), 244–250. doi:10.1207/S15324796ABM2403_10 PMID:12173682

Acknowledgment

The editors would like to acknowledge the help of all the people involved in this project. Without their support, this book would not have become a reality.

First, We would like to express our deep appreciation to each of the contributing authors for trusting us with your work. Your commitment to ensuring the success of the book means a lot to us.

Second, We wish to acknowledge the valuable contributions of the reviewers regarding the insurance of chapter quality, coherence, and presentation. Your expertise and attention to detail is appreciated beyond measure.

Third, a special note of thanks to the staff at IGI Global, whose patience and contributions throughout the whole process from inception of the initial idea to final publication have been indispensable. This book would not have been possible without the ongoing professional support from IGI Global.

Fourth, a sincere and heartfelt **THANK YOU** to the numerous individuals who contributed invaluable, tireless support and encouragement. Who would have thought a chapter "Navigating the Landscape of the STEM Professoriate" in the book *Women's Influence on Inclusion, Equity, and Diversity in STEM Fields* (IGI Global, 2019) would lead us here. These individuals reminded us of the larger purpose for the book -- on a weekly, daily, or hourly basis -- whenever it was needed.

Last but not least, We would like to thank you, the reader for providing the reason for the book. We hope you find it interesting and useful for your research and studies. More importantly, we hope it promotes thoughtful conversations around broadening participation and retention for women of color in STEM.

Pamela M. Leggett-Robinson
PLR Consulting, USA

Brandi Campbell Villa
Belay Consulting, USA

Chapter 1
Negotiating Intersectionality:
The Triple Pressure of Being an African American Female Scientist

Angela White
North Carolina A&T State University, USA

Jessica T. DeCuir-Gunby
North Carolina State University, USA

Cailisha Petty
North Carolina A&T State University, USA

ABSTRACT

This concurrent mixed methods study sought to unveil how historically Black colleges and universities assist African American women with negotiating intersectional experiences within science disciplines. The authors sought to determine the relationship(s) between racial identity, science identity, and science self-efficacy beliefs of African American science students enrolled at HBCUs. The quantitative data revealed a statistically significant positive correlation between science self-efficacy and science identity for women and men. While the quantitative data indicated that racial identity (assimilation) is positively correlated to science identity for the women, the same relationship was not significant for the men. The findings indicated that science identity significantly predicted college science achievement for men but not women.

DOI: 10.4018/978-1-7998-4858-5.ch001

INTRODUCTION

The field of science has a longstanding sociohistorical reputation and is reproduced as an "objective, privileged way of knowing pursued by an intellectual elite" (Carlone, 2004, p. 394). Science is a masculine pursuit whose members have to possess a willingness to embrace hegemonic notions of masculinity in order to achieve success (Carlone, Webb, Archer, & Taylor, 2015). Feminist researchers have shown that scientific knowledge reflects the gender and racial beliefs of society because such knowledge it is culturally situated (Brickhouse, 2001). Carlone and colleagues (2015) suggests "the 'ideal science student' is a classed, racialized, and gendered construction" (p. 438).

In comparison to African American men and White women, African American women have been consistently underrepresented in STEM. From 2012-2013, bachelor's degrees in STEM were awarded to 130,068 White men, 12,045 Black men, 64,208 White women and 9,457 Black women. This data indicates that African American women receive fewer science degrees within both their racial and gender categories. In order for our nation to remain globally competitive, there has been an aggressive push to increase the number of STEM graduates. For this reason, it is critical to give attention to how our secondary and post secondary educational systems are preparing or preventing African American females from being participants in this ever-increasing STEM workforce. While African American women receive fewer degrees in science disciplines and move into science careers at lower rates than their White counterparts and African American males, historically Black Colleges and Universities (HBCUs) have produced a disproportionately higher number of African American females than predominantly White institutions (PWIs) (Chang et al, 2008; NCES, 2017). The purpose of this chapter is to illuminate the identity negotiation process of African American women in science, and gain more insight into the role of HBCUs in such processes.

In comparison to PWIs, qualitative data suggests that HBCUs are warmer, more nurturing, community-oriented, student-centered, promote a greater sense of belonging, enhance self-efficacy beliefs, and augment racial pride (Allen, 1992; Palmer, Davis, & Maramba, 2010; Shorette & Palmer, 2015; Sloan, & Clarke, 2009; Spurgeon & Myers, 2010; Van Camp, Barden). Researchers have discovered that African American students on HBCU campuses experience a high degree of faculty interaction and possess strong sense of community (Allen, 1992; Fries-Britt & Turner, 2002; Jett, 2013; Palmer, Davis, & Maramba, 2010; Shorette & Palmer, 2015). A body of research indicates that African American students attending HBCUs experience strong peer networks that contribute to their success confidence and sense of belonging (Allen, 1992; Jett, 2013; Palmer, Davis, & Maramba, 2010). Moreover, studies indicate that attending HBCUs positively influences a student's self-concept (Shorette & Palmer, 2015; Van Camp, Barden, Sloan, & Clarke, 2009). Van Camp et al. (2009) found that many students attend HBCUs due to a desire to develop their racial identity.

African American Women and Women of Color in Science

The literature exploring African American women in science is sparse (Ong et al., 2011). According to Ong and colleagues (2011), since the turn of the century, there has been an increase in research that focuses on women of color in STEM, with African American women as the most researched group. Because of the lack of literature, it was not always possible to find articles that focused on science alone. Additionally, in many of these articles, African American women are often studied with other women of color.

Statistics

Women of color have been awarded more stem degrees since the 1970s, but are consistently underrepresented in advanced education and career stages relative to white women and men of color. Proportionally, women of color are underrepresented relative to their representation in the U.S. population compared to white and Asian American/Pacific Islander women. For example, in 2006, African American women compose 6.54% of the population (ages 15–24), but were only 5.31% of the STEM bachelor's degrees. At the doctoral level, African American women represent 6.01% of the population (ages 25–44), but represent only 2.61% of STEM PhD recipients.

Clearly, the U.S. education system and research infrastructures systematically undereducated and underutilize women of color. The daunting magnitude of their underrepresentation in advanced STEM areas represents serious equity concerns that connects with important historical and contemporary issues of social justice in the U.S. education and employment systems. (Ong et al., 2011, p. 175)

Women or minorities are usually targets for efforts in STEM. There have been programs created with the intention of increasing the pool of minorities and women in science. "Unfortunately, history has borne out the reality that programs intended to serve women disproportionately benefit White women, and programs intended to serve minorities mainly benefit minority males" (Ong et al., 2011, pp. 175–176). When focus and resources are put in place to address the needs of women or minority populations, there is a disregard for the double bind that women of color must contend with. This double bind takes into account how race/ethnicity and gender function simultaneously to produce distinct experiences (Ong et al., 2011).

Because of the lack of research on women of color in STEM, there is a lack of coherent knowledge. "A dedicated research base about women of color would help assess the root causes of attrition, retention, or advancement for this population; to identify and remedy gaps in the research; and to broadly examine and improve programmatic, institutional, and nationwide efforts" (Ong et al., 2011, p. 176). Our current knowledge base is attributed to the researchers that have privileged the experience of women of color in science.

The number of women of color in science and engineering has increased overall within the last several decades. Many scholars point to the social and structural environments in college as the main source of women of color's attrition in graduate STEM education (Carlone & Johnson, 2007; Hanson, 2004; Ong, 2005; Ong et al., 2011). Some of the key factors that influence women's development of identities in science include the undergraduate learning environment, STEM enrichment programs, relationships and self-concepts. Several studies have illustrated that supportive environments in STEM are found in minority serving institutions, especially historically Black colleges and universities (HBCUs). The key factors in these institutions are openness, lack of stigma when remediation is required, high expectations, and supportive faculty student relationships (Ong et al., 2011). Women of color valued environments where the instructors' pedagogical approach fostered cooperation instead of competition and included various opportunities for class participation and engagement.

THEORETICAL FRAMEWORK: CRITICAL RACE THEORY

This study utilizes Critical Race Theory (CRT) as a lens to explore the relationship between racial identity and science identity. Thus, race was positioned at the nucleus of analysis. CRT is a movement by those who are committed to "studying and transforming the relationship between race, racism, and power" (Delgado & Stefancic, 2001, p. 2). The movement began within the field of law, but has spread well beyond the boundaries of the courtroom (Delgado & Stefancic, 2001). CRT is about not just understanding and critiquing, but is ultimately dedicated to transformation. Therefore, CRT has moved into classrooms and crossed boundaries into other disciplines (Delgado & Stefancic, 2001; Ladson-Billings, 2009). The utilization of CRT requires that one consider that race and racism intersect with various dimensions of a person's identity such as gender, sexuality, and social class (Crenshaw, 2011).

Delgado and Stefancic (2017) point to the following central tenets of CRT: (1) permanence of racism, (2) a critique of liberalism, (3) the use of storytelling/counter storytelling – voices of people of color, (4) structural determinism (5) interest convergence and (6) intersectionality with regard to race, sex, and class. Three of the CRT tenets strongly guided this study (see Table 1 for description of

CRT central tenets). Thus, a detailed discussion has been dedicated to the following tenets: intersectionality, permanence of racism, and whiteness as property.

Intersectionality

Kimberlé Crenshaw (1995) introduced the CRT tenet of Intersectionality in an effort to shed light on the oppression and violence uniquely experienced by women of color. According to Crenshaw (2016), intersectionality should not be reduced to a notion of multiple identities. Intersectionality is more accurately operationalized as a way of exploring how power structures and systems make certain identities the "consequence of and vehicle for vulnerability" (Crenshaw, 2016).

Intersectionality draws attention to how the unique gender and race discrimination experienced by Black women was buried under the gender experiences of White women and the race experiences of Black men (Crenshaw, 1989/2016). To this end, experiences of Black women should not be lumped in with the "women's experiences" or the "Black experience" but evaluated utilizing an Intersectional framework. By not identifying the multiple, overlapping, intersecting forms of oppression, Black women will never be fully empowered (Collins, 2000; Parsons & Mensah, 2010). We must also consider the degree to which the intersections of Black women's racial, gender, and scientific identities conflict with the messages Black women and girls receive throughout their educational journey, because this may hinder their ability to successfully develop a Black woman scientist identity (Charleston, L., Lang, N., Adserias, R., & Jackson, J., 2014).

Permanence of Racism

CRT asserts that race is socially constructed and is permanently woven into the fabric of our nation (Delgado & Stefancic, 2012; Dixson, 2006; Gotanda, 1991; Ladson-Billings & Tate, 1996). One of the most salient tenets of CRT is the permanence of racism (Bell, 1992; 1995). According to Derrick Bell (1992, 1995), Black people will never experience equality in America because racism is endemic and permanent in America. From the civil rights movement to Black Lives Matter, African Americans have experienced some advancement; however, it is in no way equal to white Americans. In spite of the advancement of African Americans, negative stereotypes associated with African Americans are still prevalent and the white social norm still permeates popular consciousness (Crenshaw, 1988). To this end, Bell (1995) posits that racial equality is not a realistic goal, thus one must consider the role racism has played in America [past and present].

Table 1. CRT Tenets: Descriptions and Literature sources

CRT Tenet	Description	Literature Sources
Permanence of racism	The notion that racism is a normal part of American society – racism is ordinary, not aberrational	Bell, 1992,1995; Crenshaw, 1988; Lawrence, 1995; Peller, 1995
Challenge to Dominant Ideology	Criticizes or challenges claims of meritocracy in society, the colorblind paradigm, incremental change, and neutrality law	Cook, 1990; Crenshaw, 1988; Freeman, 1995; Gotanda, 1991; Delgado & Stefancic, 2012; Yosso, Solorzano & Parker, 2002
Voices of people of color	The lived racialized experiences of people of color are captured through storytelling/counter-storytelling; aims to cast doubt on the validity of widely accepted myths/messages (in particular those held by the majority)	Bonilla-Silva, 2002; Cook, 1990; Delgado & Stefancic, 2012; Matsuda, 1991
Structural determinism	Widely shared practice dictates significant social outcomes; Due to the structure of certain systems some problems will not be resolved	Delgado & Stefancic, 2012
Interest convergence or material determinism	The interests of people of color will only be granted when they converge with the interests of Whites – mutually beneficial. The Fourteenth Amendment does not provoke a judicial solution leading to effective racial equality for blacks when the solution poses a threat to the superior social status of whites Example: Brown v. Board of Education	Bell, 1980, 1995; Delgado & Stefancic, 2012
Intersectionality	Explores the intersecting roles of race, sexuality, gender, class	Crenshaw, 2011; Gilborn, 2015;
Whiteness as property	The law is a contributor to the construction of race. The US court system drew boundary lines around "whiteness"; therefore, Whiteness is a form of property that is protected by American law. Whiteness presented its owners with facets of citizenship, which became valuable because they were denied to those who were not identified as white. Property functions of whiteness include: right to transfer, the right to use or enjoyment, and the right to include or exclude others.	DeCuir and Dixson, 2004; Haney-Lopez, 2006; Harris, 1995; Leong, 2013

Whiteness as Property

The final tenet of CRT guiding this study is Whiteness as property. Early CRT scholars provided evidence that the U.S. judicial system reified the construction of race and perpetuated racism (Haney-Lopez, 2006; Harris, 1993/1995). The laws established by the US judicial system enforced differential treatment based on racial categories in that black people were not allowed to travel without permits, own property or acquire knowledge through formal education (Harris, 1993/1995). For this reason, the courts developed criteria by which they placed the boundaries around "whiteness" (Haney-Lopez, 2006). Harris (1993/1995) developed the CRT tenet of whitenesss as property based on the larger historical conceptualization of property, in that, property is anything that has been given a value and to which individuals have a right. The property functions of whiteness include: the right to transfer, the right to use or enjoyment, and the right to include or exclude (Harris, 1993/1995). For example, Mexican Americans were granted "white" status in 1848, but in 1930, they were taken out of the white category and given their own separate identifier. The reclassification is evidence of the diminishing status of Mexicans in the U.S. In essence, they were asked to turn in their "white card." Not having white status meant, and still means, "limited and structurally (informally and formally) circumscribed political, economic, and educational influence" (Parsons, 2014, p. 700).

INTERSECTIONALITY OF RACIAL IDENTITY AND SCIENCE IDENTITY DEVELOPMENT OF AFRICAN AMERICAN STUDENTS

The manner in which individuals experience their racial identity is interwoven with other aspects of their identity (Goodman & Jackson III, 2012). In fact, racial identity scholars are modifying or developing conceptual models that depict the complex interrelationship between race and other social identities, while integrating aspects of what has been theorized as Intersectionality (Goodman & Jackson III, 2012). The present study was shaped by previous studies that propose Students of Color perceive ethnic/racial and academic (science) identities as relevant (Syed, 2010; Walker & Syed, 2013). According to Walker and Syed's study (2013), Students of Color demonstrate greater ethnic-academic identity integration than their White counterparts; thus, they endorse higher degrees of integration between their ethnic background and their academic major. In other words, a Student of Color negotiate to who they are and how they identify ethnically and culturally and who they are becoming as scientist. Will they have to negotiate parts of who they are culturally to be a scientist? Getting recognized as a "science person" often clashes with cultural identity because very often minority students must struggle with issues of being considered a "science geek" or a "science nerd" (Hurtado et al., 2009; Rao & Flores, 2007). White male students do not have the same experience because the normative ways of speaking, behaving, and performing as a scientist have been established by the dominant culture. In the following sections, we will explore the concepts of Black racial identify followed by Science identity.

Racial Identity

Black racial identity (BRI) has been described as "the attitudes and beliefs that an African American has about his or her belonging to the Black race individually, the Black race collectively, and their perceptions of other racial groups" (DeCuir-Gunby, 2009, p. 103). Historically, theorists have developed BRI models based on two lines of thinking: process or content (Scottham, Cooke, & Ford, 2010). Theorists who utilize a process lens focus on the manner in which attitudes regarding race develop and transform across the lifespan (Scottham, Cooke, & Ford, 2010). Conversely, those who utilize a content lens place greater emphasis on the substance of one's attitudes toward this racial membership (Scottham, Cooke, & Ford, 2010). With regards to the present study, emphasis was placed on the content of students' race-related attitudes and beliefs.

Multidimensional Model of Racial Identity

In order to explore the content of students' attitudes toward their membership within their racial group, the Multidimensional Model of Racial Identity (MMRI) was heavily integrated into the quantitative component and influenced the development of several interview questions. The Multidimensional Model of Racial Identity was developed by Sellers, Smith, Shelton, Rowley, & Chavous (1998) to gain a deeper understanding of the meaning African Americans place on being members of that racial category. The MMRI model assumes that (1) identity is both situationally influenced and a stable component of the individual, (2) individuals have multiple identities that are arranged in a hierarchy and (3) the person's perception of their racial identity is the very best indicator of his identity. This multidimensional model explores three stable identity dimensions: racial centrality, racial regard, and racial ideology (Sellers, Smith, Shelton, Rowley, & Chavous, 1998). While racial centrality measures the degree by which individuals connect to race when defining themselves, racial regard and ideology refer to the meaning individuals give to being Black (Sellers, Smith, Shelton, Rowley, & Chavous, 1998).

The MMRI was operationalized in the form of the Multidimensional Inventory of Black Identity (MIBI) (Scottham, Cooke, Sellers, & Ford, 2010; Sellers, Smith, Shelton, Rowley, & Chavous, 1998). The MIBI consists of three scales that are aligned with the three stable identity dimension: centrality, ideology, and regard (See Table 2 for sample MIBI items). The centrality scale measures the degree to which race is important to the person. There are four ideology subscales that make up the MIBI: Humanist, Assimilationist, Oppressed minority, and Nationalist; however, only Nationalist and Assimilationist were used in this study. Last, the MIBI consists of two Regard subscale: public regard and private regard; however, only the public regard subscale was used in this study. Previous studies (Jones & Day, 2018) have extended this multidimensional theory using an intersectional framework. Furthermore, Jones and Day's (2018) findings suggest that identities can be hierarchical (as theorized by the MMRI) or they may be intersectional. With regards to race and gender, these two identities may be equally important or intersectional or ranked, in that one is seen as more salient than the other.

Table 2. Sample Items from MIBI subscales: Centrality, Public Regard, Assimilationist, and Nationalist

Subscale	Sample Items
Centrality	• Being Black is an important reflection of who I am. • In general, being Black is an important part of my self-image
Public Regard	• Overall, Blacks are considered good by others. • In general, others respect Black people.
Assimilationist	• A sign of progress is that Blacks are in the mainstream of America more than ever before • Because America is predominantly White, it important that Blacks go to White schools that they can gain experience interacting with White
Nationalist	• It is important for Black people to surround their children with Black art, music, and literature. • Black students are better off going to schools that are controlled and organized by Blacks

Science Identity

Within the last decade the development of a strong science identity has been recognized as a key ingredient to learning scientific concepts (Brickhouse, 2001; Carlone & Johnson, 2007). Science identity may be described as who students are, what they are capable of doing, and what they aspire to do and become as it relates to science (Brickhouse, 2001). More specifically, students' science identity is based on their perceptions of themselves and how they perceive others view them as they participate in science-related tasks (Aschbacher, Li, & Roth, 2010).

It has been very challenging for researchers to fully operationalize science identity; however, Carlone and Johnson (2007) devised a model based on their ethnographic study of women of color. Their model suggests three factors that influence the development of science identity: recognition, performance, competence. Carlone and Johnson (2007) explains the three categories as follows:

1. Performance: social performances of relevant scientific practices (i.e. ways of talking and using tools)
2. Recognition: recognizing oneself and getting recognized by others as a "science person"
3. Competence: knowledge and understanding of science content (may be less publicly visible than performance).

The Role of Historically Black Colleges and Universities (HBCUs)

The Higher Education Act of 1965 defines an HBCU as "…any historically black college or university that was established prior to 1964, whose primary mission was, and is, the education of black Americans" (National Center for Education Statistics, 2006). While HBCUs only make up a small percentage of the nation's higher education institutions, they graduate nearly twenty percent of African Americans who earn undergraduate degrees (National Center for Education Statistics, 2006). HBCUs have produced a disproportionately high number of African American graduates in STEM (Shorette & Palmer, 2015), and are successful in providing African American students with a stronger sense of self (Fries-Britt & Turner, 2002; Van Camp, Barden, Sloan, & Clarke, 2009). Of the top 20 institutions awarding science and engineering degrees to African Americans between 2008-2012, eight of the institutions were HBCUs (NSF, 2015).

In comparison to PWIs, qualitative data suggests that HBCUs are warmer, more nurturing, community-oriented, student-centered, promote a greater sense of belonging, enhance self-efficacy beliefs, and augment racial pride (Allen, 1992; Palmer, Davis, & Maramba, 2010; Shorette & Palmer, 2015; Sloan, & Clarke, 2009; Spurgeon & Myers, 2010; Van Camp, Barden). Researchers have discovered that African American students on HBCU campuses experience a high degree of faculty interaction and possess strong sense of community (Allen, 1992; Fries-Britt & Turner, 2002; Jett, 2013; Palmer, Davis, & Maramba, 2010; Shorette & Palmer, 2015). A body of research indicates that African American students attending HBCUs experience strong peer networks that contribute to their success confidence and sense of belonging (Allen, 1992; Jett, 2013; Palmer, Davis, & Maramba, 2010). Moreover, studies indicate that attending HBCUs positively influences a student's self-concept (Shorette & Palmer, 2015; Van Camp, Barden, Sloan, & Clarke, 2009). Van Camp et al. (2009) found that many students attend HBCUs due to a desire to develop their racial identity.

Methods

A Critical Race Mixed Methodology (CRMM) was employed to conduct this research. CRMM is "the combining of Critical Race Theory and mixed methods research" (DeCuir-Gunby & Schutz, 2018, p. 166). The mixed methods dimension was a concurrent mixed methods design (QUANT + qual), as this design allowed us to collect various forms of data simultaneously, analyze data separately, and subsequently combine the research findings (DeCuir-Gunby & Schutz, 2017). More attention was dedicated to the quantitative aspect of the study, while the qualitative aspect was used to corroborate the quantitative findings (Creswell & Clark, 2011); thereby, providing a richer, fuller understanding of the relationships between the racial identity and science identity of African American female students. Due to the difference in the degree of attention given to each aspect of the study, the authors notate the component with the greatest amount of attention with all capital letters, while the component with less emphasis is represented using all lowercase letters. Therefore, the use of "QUANT" indicates the stronger emphasis on the quantitative aspect, whereas "qual" represents the lesser emphasis placed on the qualitative aspect. Additionally, the "+" symbol indicates the simultaneous/concurrent data collection of both data sets (Creswell & Plano Clark, 2011).

The methodological approach for this study included: (1) collecting quantitative and qualitative data concurrently; (2) analyzing the quantitative and qualitative data separately; (3) integrating the findings from the two data sets; and (4) interpreting to what extent the quantitative and qualitative data converge to provide a fuller understanding of the problem (DeCuir-Gunby & Schutz, 2017). The authors addressed the following research questions:

1. What is the relationship between the racial identity and science identity of African American female students attending HBCUs?
2. What is the role of the HBCU context in African American females' process of negotiating the intersection of gender identity, racial identity, and science identity?

Participant Descriptions

The participants (n = 191) for the quantitative aspect of this study were self-identified African American female undergraduate students from five historically Black colleges and universities (HBCUs) located in the southeastern United States. The science majors represented by the participants included: biology, chemistry, physics and pharmaceutical sciences. Additionally, participants were sophomores, juniors, or seniors at their home institution. Table 3 provides descriptive statistics/characteristics of students who participated in the quantitative component. The qualitative aspect involved nine females from four of the five institutions, and each of them participated in the quantitative component.

Table 3. Characteristics of Student Participants from 5 Selected HBCUs in North Carolina

Variable	Female (191)	
	Mean	SD
Age	21.69	4.75
HS GPA	3.40	0.51
Science Family (#)	2.06	0.97
Science Course in HS (#)	3.32	0.81
Mathematics Courses in HS (#)	3.33	0.56
Science Courses in College (#)	8.31	4.64
Mathematics Course in College (#)	3.95	1.90
Science Program Intern (#)	0.92	1.14
Science Self Efficacy	6.23	0.65
Assimilation	4.87	0.87
Centrality	4.92	1.23
Nationalist	4.14	1.02
Public Regard	3.66	0.94
Science Identity	3.11	0.66
College GPA	3.12	0.50
Science College GPA	3.01	0.55

Participants' Institutions

The participants in this study were enrolled in one of five HBCUs located in a southeastern state. To maintain the anonymity of the participating institutions, pseudonyms were assigned to each institution; therefore, they will be referred to as Franklin State University, Jackson B. Southerland University, Smith University, North Central M&T State University, and Calvin University. While all of the institutions have biology and chemistry programs, not all of them have physics programs. Additionally, only one had a pharmaceutical science program.

Instrumentation

Several instruments were used to explore the research questions and hypotheses for this study. Students who participated in the quantitative component responded to questions on three different instruments: Multidimensional Inventory of Black Identity (MIBI) (Sellers, Smith, Shelton, Rowley, & Chavous, 1998), Science Identity Scale (Chemers et al., 2011), and Science Self-efficacy Inventory (adapted from Nietfeld, Cao, & Osborne, 2006; Nietfeld, Shores, & Hoffman, 2014). Additionally, students completed a demographics questionnaire. An interview protocol was developed and followed for the qualitative aspect.

Multidimensional Inventory of Black Identity (MIBI)

In order to investigate the racial identity development of the participants, the authors used the Multidimensional Inventory of Black Identity (MIBI), an operationalization of the Multidimensional Model of Racial Identity (MMRI) (Sellers, Smith, Shelton, Rowley, & Chavous, 1998). This scale was developed to explore the multidimensional nature of Black Identity (Sellers, Smith, Shelton, Rowley, & Chavous, 1998). The MIBI measures three of the four dimensions of racial identity proposed by the MMRI: racial centrality, racial regard, and racial ideology (Sellers, Smith, Shelton, Rowley, & Chavous, 1998). However, it does not measure racial salience, as it is considered to be situationally-influenced rather than stable like the others (Sellers, Smith, Shelton, Rowley, & Chavous, 1998). To elaborate, racial centrality references how individuals normatively define themselves in terms of race across all contexts, and indicates that being Black/African American is an important component of one's total makeup. Racial centrality has been examined as a predictor of the ways an individual may perceive or psychologically make sense of racial experiences. This identity dimension is critical to the present study because it illuminates how African America students may perceive their race influences their ability to be a scientist and perform science-related tasks. Racial ideology captures an individual's thoughts and attitudes regarding how they think other Black people should act across all contexts (Sellers, Smith, Shelton, Rowley, & Chavous, 1998). There are four components of this dimension: Nationalist Ideology, Oppressed Minority Ideology, Assimilationist Ideology, and Humanist Ideology. For the purpose of our study, attention was given to the Nationalist and Assimilationist Ideologies. The Nationalist emphasizes the uniqueness of Black individuals, and preferences for Black spaces. The Assimilationist highlights the similarities between Blacks and White American society. Racial regard is an individual's negative or positive perception of Blacks and their understanding of others' judgment of Blacks (Sellers, Smith, Shelton, Rowley, & Chavous, 1998). There are two components of this identity dimension: Public and Private. Public regard indicates how African Americans internalize feelings about being African American, as well as how they perceive White people view Black/African American people (Sellers et al., 1998). Private regard is the extent to which one feels positively or negatively about his/her race and other Blacks (Sellers et al., 1998).

Seven subscales are used to represent each of these dimensions: Centrality, Private and Public Regard, Nationalist Ideology, Oppressed Minority Ideology, Assimilationist Ideology, and Humanist Ideology (Sellers et al., 1998). The MIBI consists of 54 items featuring a 7-point Likert format. For the purpose of this study, the authors only examined four dimensions of racial identity based on the MIBI: centrality, public regard, assimilationist ideology, and nationalist ideology. The centrality scale measures the degree to which race is important to an individual. For example, an African American female science student with a high centrality score perceives race is a very significant component of her self-concept. Additionally, in the context of this study, she would also feel very connected to her HBCU. The Public Regard subscale measures the extent to which a person feels others view African Americans positively or negatively (Sellers, Rowley, Chavous, Shelton, & Smith, 1997); therefore, a higher score in this study would indicate a female perceives white people hold more positive feelings toward African Americans while lower scores indicate a female perceives that white people hold more negative views towards African Americans. The authors elected to include the public regard subscale for this study due to the prevailing deficit messages dominant culture have perpetuated about science and mathematics abilities of African Americans. Additionally, the inclusion of the public regard subscale was included based on the stereotypical images and messages that portray science as a Eurocentric field designated for males. Therefore, the inclusion of this subscale provided insight into how such judgments inform or interfere with the development of the participants' science identity. The Assimilationist and Nationalist subscales center on racial ideology, and were utilized in our study based on previous findings indicating that Nationalist ideologies were higher than Assimilationist ideologies for students at HBCUs (Sellers et al., 1997; Cokely & Helm, 2001). Additionally, earlier studies reported higher Nationalist scores for African American students who attend HBCUs compared to those who attend PWIs, while the Assimilationist attitude was higher for those who attend PWIs. A female with a high assimilationist score would focus on the similarities between African Americans and the rest of American society. She would not necessarily de-emphasize of reject her Blackness, but believe that African Americans should work within mainstream science communities of practices to make a change. For instance, a female participant with a high assimilationist score may think she should only present her research at predominantly white science conferences or attend a predominantly white medical school. A female with a high nationalist score would focus on the uniqueness of the Black experience and maintain that uniqueness within mainstream science communities of practice or only desire to perform science within Black science communities.

With regards to reliability, Sellers et al. (1997) reported the following Chronbach's alpha for students attending HBCUs: centrality (0.75), assimilation (0.74), and humanist (0.69). In this study Chronbach's alphas were found and indicated good reliability; however, the Chronbach's alpha for the Assimilation scale is not as strong. The Chronbach's alphas for this study are as follows: Centrality (0.78), Nationalist (0.74), Public Regard (0.79), and Assimilation (0.66). The Cronbach's alpha coefficient of 0.66, which indicated questionable reliability for this study; however, this is consistent with the Cronbach's alphas reported by Sellers et al. (1997). A factor analysis indicated that a four-factored solution explained 39.85% of the variance.

Science Identity Scale

In order to gain insight into science identity, Chemers et al. (2011) developed the Identity as a Scientist scale. The development of this scale was based on Sellers' Multidimensional Model of Racial Identity/Multidimensional Inventory of Black Identity, the work of Luhtanen and Crocker (1992), and interviews. The most recent iteration of the scale is a 4-point Likert format (1 = strongly disagree and 4 = strongly agree) (Robnett, Chemers, & Zurbriggen, 2015). The reported Chronbach's alpha for internal consistency reliability for this scale ranges from 0.89 to 0.96 (Chemers et al., 2011; Robnett, Chemers, & Zurbriggen, 2015). For our sample, the Chronbach's alpha was 0.85 for Science Identity. A factor analysis was conducted, and a one-factored solution explained 53.93% of the variance.

Demographics Questionnaire

Students were asked to complete a demographic questionnaire that included questions regarding personal characteristics such as age, gender, major, and classification.

The questionnaire also captured their educational experiences such as number of high school science/mathematics courses, rigor of high school science/mathematics courses (i.e. college preparatory, advanced placement, international baccalaureate, etc.), college science/mathematics courses, SAT/ACT MATH scores, participation in science camps/programs, high school grade point average (GPA), and college GPA.

Interview Protocol

In order to learn more about the way(s) HBCUs assist their female science students with negotiating the intersection of their gender identity, racial identity, and science identity convenience sampling was employed to recruit the interview participants (n = 9) for the qualitative aspect. In order to ensure that all of the essential topics were addressed, semi-structured interviews were conducted using an interview guide (Creswell, 2007); however, probing questions were asked based on the students' responses to the questions on the protocol.

FINDINGS

The larger goal of this study was to learn more about the science identity development of African American females attending HBCUs, as well how it intersects with multiple other identities (i.e. gender and race). To determine the relationship between their science and racial identities, a Pearson correlation analysis was conducted among science identity, centrality, public regard, assimilationist, and nationalist (see Table 4). This analysis revealed that assimilationist is marginally correlated with science identity ($r = 0.17$, $p < 0.05$); therefore, as female HBCU students endorsed an assimilationist attitude their science identity increased. Additionally, this sample of females had a higher average assimilation attitude ($M = 4.87$; $SD = 0.87$) than nationalist attitude ($M = 4.14$; $SD = 1.02$). Based on this novel finding, many African American females attending HBCUs potentially adopt an attitude that would influence them to function more like the scientists they see in the rest of society. At this point, attention should be given to whether or not African American females, even those who attend HBCUs, feel they must engage in science according to the "cultural norms of science", that being the standard ways of speaking, performing experiments, learning and working in science. We must pause to consider if the intersection of an assimilationist attitude with science identity marginalizes the potentially unique contributions African American women could make to science. What is the role of HBCUs in this negotiation process?

In order to gain a richer and fuller understanding of the role of the HBCU context in African American females' negotiation of their gender, science identity,

Table 4. Correlation between Science Identity and Other Variables for Each Gender

Variable	Female		
	Corr	P-Value	n
Age	-0.06	0.3910	185
HS GPA	-0.12	0.1009	184
Science Family (#)	-0.12	0.0984	187
Science Course in HS (#)	0.00	0.9636	186
Mathematics Courses in HS (#)	-0.02	0.7739	187
Science Courses in College (#)	-0.10	0.1775	181
Mathematics Course in College (#)	-0.01	0.9109	186
Science Program Intern (#)	0.11	0.1623	174
Science Self Efficacy	0.29	0.0001	175
Assimilation	0.17	0.0279	176
Centrality	0.07	0.3655	183
Nationalist	0.05	0.4929	182
Public Regard	0.09	0.2214	181
College GPA	0.07	0.3159	187
Science College GPA	0.06	0.4353	187

and racial identity (specifically assimilation ideology), the authors conducted a thorough analysis of the interview transcripts. Since racial ideology informs an individual's perception of what she should do, the authors looked for who the participants sought to do science with, their preferences for settings to do science, and how they engaged in science.

Visible Yet Invisible

Each of the women spoke of different African American female scientists who played a critical role in their identity negotiation process. In some instances, these young women identified with and experienced mentorship from a faculty member at their home institution; however, for some the mentorship resulted from engagement in science communities of practice outside of their HBCU. In the instances where mentorship was established beyond the boundaries of the HBCU setting, the women were encouraged by faculty within their department to participate in external research or predominantly Black conferences and summer enrichment programs. To some extent, these female scientists/mentors recognized each of the young women as a scientists or catalyzed the process of the young women self-recognizing themselves as scientists. Such recognition was also accompanied by acknowledging aspects of who they were as women and as African Americans. According to Nasir (), recognition is a "relational resource" that promotes the identity development of racialized individuals. To elaborate, relational resources are the manner in which positive relationships with others in the context increase an individual's connection to a practice.

In the first example from Patricia, a junior biology student at Calvin University, identified across the three dimensions with a well known female microbiologist in her department:

It makes me inspired to continue on because I see that they did it. For example, my first biology teacher here [Calvin University] was a Black woman and she's also a scientist. Like, she got her degree in microbiology. And now she currently chairs one of the research programs I am in. So, it's like a lot of things that she does is what I could see myself doing....seeing our professors, you know, seeing what we can become if we keep going. Seeing someone that looks like us. So it definitely inspires me as a scientist.

Note the components of intersectionality that inspired Patricia to continue to develop into a science person: the biology professor was black and she was a woman; therefore, Patricia believed she too had access to science. "Seeing" African American female faculty move beyond racial and gender oppression gave them confidence that they are able to do the same. Also, within this passage, notice Patricia's self-recognition as a recognition source to her science identity development.

In the second example, Tiffany, a sophomore biology student a Jackson B. Southerland University, found that African American female faculty understood the unique struggles experienced by African American female science students. Having access to an established African American female scientists who did not base her ability to come out of challenging moments with learning science on her race or gender allowed Tiffany to continue to see herself as a science person. What follows is a specific account of how Tiffany was able to endure a challenging moment in a chemistry class without succumbing to gender or racial stereotypes which was expressed during a different portion of her interview:

One of my favorite professors here is my organic chemistry professor, Dr. Bronson. I remember one time where I just felt I couldn't do it. Everything was just getting so hard and... she was understanding the struggle and she said you're in the struggle now for this moment, but if you let this moment stop you, then 5-10 years later you're going to look back on this moment and regret that you let that stop you.

For some of the participants, African American female faculty role models were invisible within their department, specifically within the quantitative sciences. Hannah, a senior physics student at North Central M&T State University, expressed that not only are there no African American female faculty in her physics department at North

Central M&T, but there is only a small representation in the entire physics domain. For this reason, she felt pressure to adapt to the culture of the physics community and de-emphasize aspects of her gender.

It's interesting because even with being a woman in the field is very different. Because women are seen as being emotional and all this. I'm not an extreme outright person, but if something excites me like my research when I enjoy doing it and talking about it, I'm going to show that. That's one thing that a lot of physicists are emotionless. Sitting in a conference is monotone. I hate that. It's not easy because there's not a lot of black female physicists.

Hannah's identity negotiation experience is aligned with findings from other studies that investigated the masculine nature of physics and recognition (Avraamidou, 2019; Danielsson, 2012; Gonsalves, 2014). Gonsalves (2014) explored the identities of 11 men and women physics doctoral students, and found that recognition was very critical to the development of their physics identity. Additionally, the study revealed that some women compromised their femininities and performing gender neutrality performance in order to fit into the dominant culture of their physics department. Hannah's account along with Gonsalves' findings reveal the problem of the construction of physics as a discipline that is tolerant of a range of expressions of masculinities, but allows for very little expression of femininity (Danielsson, 2012).

Hannah's department consisted completely of men, with the majority being white men. Since the department strongly encouraged their students to participate in research and conferences, Hannah was able to find female mentorship when she attended various conferences. Being able to connect with these women made it possible for her to integrate her science, racial and gender identities:

It's not easy because there's not a lot of black female physicists, but luckily I have had a change to meet a lot [at conferences] and they are like my mentors now. If anything happens or any situation, I've had to reach out to them and they're like "we understand. We're here. There is a group of us. You're not alone", so that's really good.

Note the expression "we understand", thus implying that there are unique experiences within the science (in this case physics) shared by African American women that cannot be understood by White women or Black men.

Fortunately, Hannah was able to interact with Black female physicists due to her participation in various conferences; however, her cohort Shannon, also a senior physics student at North Central M&T State University, was not as fortunate. For this reason, Shannon was not able to fully integrate her science identity with her racial and gender identities. The following example explains how the invisibility of African American female faculty interfered with Shannon's science identity. Additionally, Shannon's comment revealed how the lack of recognition be the male faculty (primarily white male) as indicated by the phrase "they kind of glaze over you" disrupted the development of her science identity.

I know it's like 2017 but it's still the same, even in class I notice like I'll say something and they're [white male faculty] like no that's not it but then a guy says something and they're like that's right, when he just said literally what I just said. I don't know, it's just weird being around a whole bunch of men and two girls [she and Hannah]. Maybe I've kind of pushed it to the back since it's been four years. It's like whatever now but I noticed men look at each other when they speak and they kind of glaze over you. And then they'll look at you, then they'll glaze over you one. Or one time they have me doing shopping stuff. I don't want to shop... Like it's just weird but yeah there's way more men out there in physics for sure. Way more African American men than women out there.

In some instances, mentorship did not come through African American female faculty but a more senior African American female student. For instance, Mary, a junior chemistry student at North Central M&T State University, was able to better visualize herself as an African American scientist because of interactions with an upperclassman:

One of my mentors last year was introduced to me by Dr. Fatima, and she's now in medical school...My mentor, she attended a program and there was students from Harvard University, from all of these Ivy League schools. People didn't really believe in her because she was black, but she proved them wrong.

In this example, Mary learned from her student mentor that her competence as a scientist will be questioned in predominantly white science communities of practice due to the intersection of race and gender. Note her expression "she proved them wrong" which revealed a coping strategy employed by her mentor when confronted with prevailing stereotypes. Mary's own negotiation of race, gender and science identity was influenced by the intersectional experiences of this more experienced and credible peer.

The experiences of these young women reveal how the access to or lack of access to African American female mentors at HBCUs impact identity negotiation processes. The relational resources available to them impacted the degree to which they connected with their science discipline, thus developing a strong science identity. In some instances the relational resource came from a relationship with an African American female faculty member or student within their HBCU. In other instances the relational resource came from a relationship with African American female scientists outside of their HBCU. In the instances where the young women had access to a mentor within their HBCU, they did not demonstrate a tendency to struggle with the integration of both their racial and gender identities with their science identity.

As these young women engaged in science communities of practice within the confines of their HBCU as well as in external conferences or programs, they were repeatedly exposed to how privilege in science is produced and reproduced in daily practices and interactions. The women in this study expressed an extreme appreciation for their HBCU and enjoyed learning science at their institution. While Mary spoke of an opportunity to further develop as a scientist within a predominantly Black summer leadership program, many of her external experiences were in predominantly white programs. Additionally, after she earned her BS in chemistry, she planned to attend a pharmacy school at a PWI instead of an HBCU. She stated "I'm actually planning to attend pharmacy school at a primarily white institution. I think it will be extremely hard for me, but I have already built that foundation at an HBCU." Her decision to participate in predominantly white programs or to attend a predominantly white pharmacy school eluded to the endorsement of an assimilationist attitude. While she felt her HBCU built a strong foundation for her to handle academic rigor, racism and sexism, her science identity developed as a result of engaging in science within white spaces. To revisit how Mary's negation process was informed by her mentor, she saw participating in white science communities of practice as an opportunity to "prove them wrong". For instance, when participating in the Medical Experience Academy, a predominantly white program, Mary excitedly expressed how presenting her research "in front of whites" caused them to shift their thoughts about her credibility as a scientist. She expressed, "They're like, hold up a second, I need to maybe step up my game. The fact that you can really get them on their toes has been something that is positive that I've seen as being an African-American scientist."

While individuals may generally be categorized as predominantly possessing one racial ideology, it is probable that most people hold a variety of ideologies that sometimes vary across areas of functioning. In the case of Mary and the other women in this study, they are actively involved in African American student organizations and expressed their most positive experiences as scientists have been at predominantly Black conferences (Nationalist), they future plans as scientists involve doing science in predominantly white settings (Assimilationist). However, the findings from the interviews suggest the endorsement of an assimilationist attitude may be a strategic interaction with the science identity in an effort to dismantle racial and gender stereotypes.

For example, Hannah shared how she felt more comfortable attending predominantly Black science conferences because she was not required to employ strategies to be considered competent by white male scientists. The following account gives insight into the dichotomy of her experiences in the two science communities of practice:

I go into National Society of Black Physicist conference and other conferences that are not racially geared. They're very different experiences. At NSBP, National Society of Black Physicist conferences, I guess it's more relaxing in a sense for me. To be in a space where it's other physicists of color and I can enjoy it more in a sense because we have this understanding whereas other conferences like SPS Society of Physics Students is just more white physicists. Your typical physicists. It's different when an older white guy comes to your poster and is asking you all these questions and it's not really like they want to know more about your research or just want to know what you know. It's like you're trying to prove yourself in a sense. Through the years I have found ways to flip the tables. That's one major difference is how you handle that one on one conversation at your poster. Different between people testing your knowledge versus actually being interested in your research. If anything it's more reassuring to me because I know myself. I know what I know. Not that I know everything, but as far as my research is concerned, I'm going to know. I'm going to be on my stuff. I'm going to know. It can be discouraging at times as far as professionals in the field are not welcoming to diversity. For me, I enjoy it and it makes me happy to do what I do. I'm not going to let someone who doesn't agree with me deter me from my path. I'm still going to be a physicist. A black woman physicist. I don't know, that's just my mentality.

Note Hannah indicated that "typical physicists" were older white men, and she only felt comfortable as an African American female physicists when she attended predominantly Black science conference. Within her interview, she mentioned that it was at predominantly Black conferences that she was able to interact with a critical mass of Black female physicists. Conversely, her bid for recognition as an African American female physicist at predominantly White conferences required her to "prove" herself and "flip the tables". The strategy of flipping the tables reveals Hannah's need to exemplify qualities of "typical physicists" in order to gain recognition as a competent scientist. She functioned according to the cultural norms of physics at white conferences; therefore, she strategically assimilated as her racial and gender identities became vulnerable to hegemonic power structures.

DISCUSSION

The experiences of African American women in science are most effectively explored through an intersectional analytic lens, as intersectionality highlights their unique and simultaneous experiences with racism and sexism. This chapter shed light on how the science identity development of African American women is informed by both their race and gender. As seen through the lived experiences of the women in this study, African American women may have strong racial and gender identities, but they may conflict with the accepted gender and cultural norms of their field; therefore, interfering with the development of a strong science identity.

The current study explored the intersection of being Black and female while pursuing a science degree at HBCUs. Specifically, we explored the relationship between the racial identity and science identity of African American female science students across five southeastern HBCU campuses. Moreover, we investigated the role HBCUs in the identity negotiation process of African American women pursuing science degrees. The quantitative findings revealed a statistically significant positive relationship between the Assimilation Ideology (Racial Identity) and Science Identity for the female HBCU participants.

It appears that HBCUs provide a space were Black males are highly represented within the faculty whereas Black females are fairly invisible. This seemingly translates into the classrooms and/or labs, in that, men are advantaged and receive more recognition during dialogue. Many of the women participants indicated that recognition was important due to their desire to convey a message to younger African American females that doing science is possible. Additionally, it was important because they had a keen awareness of the underrepresentation of African American females in science. According to Carlone and Johnson (2007), Recognition is a very salient dimension of science identity. For the female HBCU students in this study, recognition was very important; however, in most instances their desire was for others to recognize the contributions of African Americans to science and society. It is highly probable that for the women in our study, the lack of recognition of African Americans' to STEM fields and specifically the lack of recognition of African American women impacted their overall science identity development. However, in situations when they were exposed to African American women scientists who received some degree of recognition for their success in the field, the participants were able to successfully negotiate the intersection of their science identity into their other identities.

The expressed experiences of the women in the highly quantitative science fields (i.e. physics and chemistry) revealed that white privilege, specifically that rights to participate in science are transferred from the white male faculty to African American male students. Additionally, when they spoke of faculty interactions, they often shared how African or African American male faculty assisted in their journey through science. Such experiences fostered the development of the perception that a physics or chemistry identity was not readily available for them to access due to the intersection of race and gender. For them to gain access, they were required to prove themselves as competent participants.

Shannon and Hannah provided a demographic sketch of their faculty in the physics department at North Central M&T, in which African American females are invisible. In fact, there were not any female faculty, let alone an African American female, in which they could closely identify with as a scientist. This information led our team to investigate the demographic makeup of the science faculty at each of the HBCUs as an attempt to determine the representation of African American female faculty. Across all of the five HBCUs involved in this study, there were not any African American female faculty members within the physics departments. It is important to note that Smith University and Jackson B. Southerland do not have physics departments. Table 5 provides a representation of the African female faculty across the five institutions by science field. The table does not include adjunct faculty, only full-time. As shown in the table the highest representation of African American female faculty across each of the five institutions is within the field of biology, whereas the representation for chemistry and physics is relatively nonexistent. Such lack of visibility of and access to African American female faculty further reifies the notion of whiteness as property, specifically the right to exclude and include. The invisibility of African American female faculty is a huge concern from a CRT perspective, because African American female science students do not have consistent access of role models within their departments. This finding revealed that HBCUs reproduce controlling images of Black womanhood, stereotypes of science ability and the invisibility of Black women in science. The institutional structures to a degree contribute to the exclusion of some and not others, which supports the notion of Intersectionality.

Table 5. Representation of African American Female Science Faculty Across Five HBCUs

	Chemistry	Physics	Biology
North Central M&T	4 out of 12	0	13 out of 25
Calvin University	2 out of 15	0	3 out of 18
Smith University	0	No Department	3 out of 3
Franklin State University	0	0	1 out of 15
Jackson B. Southerland University	0	No Department	1 out of 5

Our unique approach of framing the study with Intersectionality and the Multidimensional Model of Racial Identity lends to the need to approach the science-related experiences of African American from a gendered racial identity lens (Thomas, Hacker, & Hoxha, 2011). Both the quantitative and qualitative findings validate the MMRI (Sellers et al., 1998), one of the theoretical underpinnings of this study. Additionally, they showcase the utility of extending the multidimensional theory through the utilization of an intersectional framework. In alignment with previous research (Jones & Day, 2018), exploring identity from a multidimensional lens promoted a revealing of differences across aspects of identity. This multidimensional and intersectional approach demonstrated that identity is not similarly constructed by all Black women. Like Jones and Day (2018) and in alignment with the assumptions of the MMRI, our results indicate that women assign varying meaning to their intersectional identity. This conclusion was supported by varying levels of centrality and meaning ascribed to being Black, specifically reflected in the assimilation finding.

This work demonstrates that it is vitally important to re-conceptualize science identity in light of other identities, particularly for African American women. There are power structures, inequities, racism, sexism, and the exclusion of certain groups from science. It is important for HBCUs to examine inequities within tenure and hiring processes that promote invisibility of African American female faculty. Their presence influences the intersectional construction of identity for African American females in science.

REFERENCES

Allen, W. (1992). The color of success: African American college student outcomes at predominantly White and historically Black public colleges and universities. *Harvard Educational Review*, *62*(1), 26–44. doi:10.17763/haer.62.1.wv5627665007v701

Aschbacher, P., Li, E., & Roth, E. (2010). Is science me? High school students' identities, participation and aspirations in science, engineering, and medicine. *Journal of Research in Science Teaching*, *47*(5), 564–582.

Avraamidou, L. (2019). "I am a young immigrant woman doing physics and on top of that I am Muslim": Identities, intersections, and negotiations. *Journal of Research in Science Teaching*.

Avraamidou, L. (2019). Science identity as a landscape of becoming: Rethinking recognition and emotions through an intersectionality lens. *Cultural Studies of Science Education*. doi:10.100711422-019-09954-7

Bandura, A. (1977a). Self-efficacy: Toward a unifying theory of behavioral change. *Psychological Review*, *84*(2), 191–215. doi:10.1037/0033-295X.84.2.191 PMID:847061

Bandura, A. (1977b). *Social learning theory*. Englewood Cliffs, NJ: Prentice Hall.

Bandura, A. (1986). *Social foundation of thought and action: A social cognitive theory*. Englewood Cliffs, NJ: Prentice Hall.

Bandura, A. (1997). *Self-efficacy: The exercise of control*. New York: Freeman.

Beier, M., Miller, L., & Wang, S. (2012). Science games and the development of scientific possible selves. *Cultural Studies of Science Education*, *7*(4), 963–978. doi:10.100711422-012-9408-0 PMID:23483731

Bell, D. (1992). *Faces at the bottom of the well: The permanence of racism*. New York: Basic Books.

Bell, D. (1995). Racial realism. In K. Crenshaw, N. Gotanda, G. Peller, & K. Thomas (Eds.), *Critical race theory: The key writings that formed the movement* (pp. 302–312). New York: The New Press.

Brickhouse, N. (2001). Embodying Science: A feminist perspective on learning. *Journal of Research in Science Teaching*, *38*(3), 282–295. doi:10.1002/1098-2736(200103)38:3<282::AID-TEA1006>3.0.CO;2-0

Brickhouse, N., Lowery, P., & Schultz, K. (2000). What kind of a girl does science? The construction of school science identities. *Journal of Research in Science Teaching*, *37*(5), 441–458. doi:10.1002/(SICI)1098-2736(200005)37:5<441::AID-TEA4>3.0.CO;2-3

Britner, S., & Pajares, F. (2006). Sources of science self-efficacy beliefs of middle school students. *Journal of Research in Science Teaching*, *43*(5), 485–499. doi:10.1002/tea.20131

Brown, B. (2004). Discursive identity: Assimilation into the culture of science and its implications for minority students. *Journal of Research in Science Teaching*, *41*(8), 810–834. doi:10.1002/tea.20228

Carlone, H. (2004). The cultural production of science in reform-based physics: Girls' access, participation, and resistance. *Journal of Research in Science Teaching*, *41*(4), 392–414. doi:10.1002/tea.20006

Carlone, H., & Johnson, A. (2007). Understanding the science experiences of successful women of color: Science identity as an analytic lens. *Journal of Research in Science Teaching*, *44*(8), 1187–1218. doi:10.1002/tea.20237

Carlone, H., Webb, A., Archer, L., & Taylor, M. (2015). What kind of boy does science? A critical perspective on the science trajectories of four scientifically talented boys. *Science Education*, *99*(3), 438–464. doi:10.1002ce.21155

Ceglie, R. (2011). Underrepresentation of women of color in the science pipeline: The construction of science identities. *Journal of Women and Minorities in Science and Engineering*, *17*(3), 271–293. doi:10.1615/JWomenMinorScienEng.2011003010

Chang, M., Cerna, O., Han, J., & Saenz, V. (2008). The contradictory role of institutional status in retaining underrepresented minorities in biomedical and behavioral science majors. *The Review of Higher Education, 31*(4), 433–464. doi:10.1353/rhe.0.0011

Chemers, M., Zurbriggen, E., Syed, M., Gaza, B., & Bearman, S. (2011). The role of efficacy and identity in science career commitment among u3nderrepresented minority students. *The Journal of Social Issues, 67*(3), 469–491. doi:10.1111/j.1540-4560.2011.01710.x

Cohen, J. (1988). *Statistical power analysis for the behavior sciences* (2nd ed.). St. Paul, MN: West Publishing Company.

Collins, P. H. (2000). *Black feminist thought.* New York, NY: Routledge.

Crenshaw, K. (1988). Race, reform, and rentrenchnment: Transformation and legitimation in anti-discrimination law. *Harvard Law Review, 101*(7), 1331–1387. doi:10.2307/1341398

Crenshaw, K. (1995). Mapping the margins: Intersectionality, identity politics, and violence against women of color. In K. Crenshaw, N. Gotanda, G. Peller, & K. Thomas (Eds.), *Critical race theory: The key writings that formed the movement* (pp. 357–383). New York: The New Press.

Crenshaw, K. (2011). Twenty years of critical race theory: Looking back to move forward. *Connecticut Law Review, 43*(5), 1253.

Crenshaw, K. (2016). *WOW Conference Keynote presentation.* Retrieved from https://www.youtube.com/watch?v=-DW4HLgYPlA

Creswell, J. (2007). *Qualitative inquiry & research design: Choosing among five approaches.* Thousand Oaks, CA: SAGE Publications.

Creswell, J., & Plano-Clark, V. (2011). *Designing and conducting mixed methods research.* Thousand Oaks, CA: Sage Publications, Inc.

Danielsson, A. (2012). Exploring woman university physics students 'doing physics' and 'doing science'. *Gender and Education, 24*(1), 25–39. doi:10.1080/09540253.2011.565040

DeCuir, J., & Dixson, A. (2004). So when it comes out, they aren't that surprised that it is. *Educational Researcher, 33*(5), 26–31. doi:10.3102/0013189X033005026

DeCuir-Gunby, J. (2009). A review of the racial identity development of African American Adolescents: The role of education. *Review of Educational Research, 79*(1), 103–124. doi:10.3102/0034654308325897

DeCuir-Gunby, J., Marshall, P., & McCulloch, A. (2011). Developing and using a codebook for the analysis of interview data: An example from a professional development research project. *Field Methods, 23*(2), 136–155. doi:10.1177/1525822X10388468

DeCuir-Gunby, J., & Schutz, P. (2017). *Developing a mixed methods proposal: A practical guide for beginning researchers*. Thousand Oaks, CA: SAGE Publications Inc. doi:10.4135/9781483399980

DeCuir-Gunby, J. T., & Schutz, P. A. (2018). Critical race mixed methodology: Designing a research study combining critical race theory and mixed methods research. In J. T. DeCuir-Gunby, T. K. Chapman, & P. A. Schutz (Eds.), *Understanding critical race research methods and methodologies: Lessons from the field* (pp. 166–179). Routledge. doi:10.4324/9781315100944-14

DeCuir-Gunby, J. T., & Walker-DeVose, D. C. (2013). Expanding the counterstory: The potential for critical race mixed methods studies in education. Handbook of critical race theory in education, 248-259.

Delgado, R., & Stefancic. (2012). *Critical race theory: An introduction* (2nd ed.). New York, NY: New York University Press.

Dixson, A. (2006). *Critical race theory in education: All God's children got a song*. New York, NY: Routledge.

Farmer, T., Elliott, S., & Eyles, J. (2006). Developing and implementing a triangulation protocol for qualitative health research. *Qualitative Health Research, 16*(3), 377–394. doi:10.1177/1049732305285708 PMID:16449687

Fetters, M., Curry, L., & Creswell, J. (2013). Achieving integration in mixed methods designs: Principles and practices. *Health Services Research, 48*(6pt2), 2134–2156. doi:10.1111/1475-6773.12117 PMID:24279835

Fraser, J., Shane-Simpson, C., & Asbell-Clarke, J. (2014). Youth science identity, science learning, and gaming experiences. *Computers in Human Behavior, 41*, 523–532. doi:10.1016/j.chb.2014.09.048

Fries-Britt, S., & Turner, B. (2002). Uneven stories: Successful Black collegians at a Black and a White campus. *The Review of Higher Education, 25*(3), 315–330. doi:10.1353/rhe.2002.0012

Gasman, M., Baez, B., & Stello Viernes, C. (2007). *Understanding minority serving institutions*. Albany, NY: SUNY Press.

Gonsalves, A. (2014). Physics and the girly girl – There is a contradiction somewhere': Doctoral students' positioning around discourses of gender and competence in physics. *Cultural Studies of Science Education, 9*(2), 503–521. doi:10.100711422-012-9447-6

Goodman, D., & Jackson, B. (2012). Pedagogical Approaches to Teaching about Racial Identity from an Intersectional Perspective. In C. Wijeyesinghe & B. Jackson (Eds.), *New Perspectives on Racial Identity Development: Integrating Emerging Frameworks* (2nd ed., pp. 216–240). NYU Press.

Gotanda, N. (1991). A critique of "Our constitution is color-blind.". *Stanford Law Review, 44*(1), 1–68. doi:10.2307/1228940

Haney-Lopez, I. (2006). *White by law: The legal construction of race*. New York, NY: New York University Press.

Hanson, S. L. (2004). African American women in science: Experiences from high school through post-secondary years and beyond. *NWSA Journal, 16*(1), 96–115. doi:10.2979/NWS.2004.16.1.96

Harris, C. (1993). Whiteness as Property. *Harvard Law Review, 106*(8), 1707–1791. doi:10.2307/1341787

Harris, C. (1995). Whiteness as Property. In K. Crenshaw, N. Gotanda, G. Peller, & K. Thomas (Eds.), *Critical race theory: The key writings that formed the movement* (pp. 276–291). New York: The New Press.

Hazari, Z., Sadler, P., & Sonnert, G. (2013). The science identity of college students: Exploring the intersection of gender, race, and ethnicity. *Journal of College Science Teaching, 42*(5), 82–91.

Hill Collins, P. (1986). Learning from the outsider within: The sociological significance of Black feminist thought. *Social Problems, 33*(6), S14–S32. doi:10.2307/800672

Hill Collins, P. (1989). The social construction of Black feminist thought. *Signs (Chicago, Ill.), 14*(4), 745–773. doi:10.1086/494543

Hill Collins, P. (2000). *Black feminist thought: Knowledge, consciousness, and the politics of empowerment* (2nd ed.). New York, NY: Routledge.

Hurtado, S., Cabrera, N. L., Lin, M. H., Arellano, L., & Espinosa, L. L. (2009). Diversifying science: Underrepresented student experiences in structured research programs. *Research in Higher Education, 50*(2), 189–224. doi:10.100711162-008-9114-7 PMID:23503690

Jett, C. (2013). HBCUs propel African American male mathematics. *Journal of African American Studies, 17*(2), 189–205. doi:10.100712111-011-9194-x

Jones, M., & Day, S. (2018). An exploration of Black women's gendered racial identity using a multidimensional and intersectional approach. *Sex Roles, 79*(1-2), 1–15. doi:10.100711199-017-0854-8

King. (1985). Multiple jeopardy, multiple consciousness: The context of a Black feminist ideology. In Guy-Sheftall (Ed.), *Words of fire: An anthology of African American feminist thought*. New York: The New Press.

Ladson-Billings, G. (2009). Just what is Critical Race Theory and what's it doing in a nice field like education? In E. Taylor, D. Gillborn, & G. Ladson-Billings (Eds.), *Foundations of Critical Race Theory in education* (pp. 17–36). New York, NY: Routledge.

Ladson-Billings, G., & Tate, W. (1996). Toward a critical race theory of education. *Teachers College Record, 97*(1), 47–68.

Lave, J., & Wenger, E. (1991). Situated learning: Legitimate peripheral participation. New York, NY: Cambridge.

Leslie, S. J., Cimpian, A., Meyer, M., & Freeland, E. (2015). Expectations of brilliance underlie gender distributions across academic disciplines. *Science, 347*(6219), 262–265. doi:10.1126cience.1261375 PMID:25593183

Luhtanen, R., & Crocker, J. (1992). A collective self-esteem scale: Self-evaluation of one's social identity. *Personality and Social Psychology Bulletin, 18*(3), 302–318. doi:10.1177/0146167292183006

Malone, K., & Barabino, G. (2009). Narrations of race in STEM research settings: Identity formation and its discontents. *Sci. Ed., 93*(3), 485–510. doi:10.1002ce.20307

Margolis, J., Fisher, A., & Miller, F. (2000). An anatomy of interest: Women in undergraduate computer science. *Women's Studies Quarterly, 28*, 104–127.

Matsuda, M. (1995). *Looking to the bottom: Critical legal studies and reparations. In Writings that formed the movement* (pp. 63–79). New York: The New Press.

Nasir, N. S., McKinney de Royston, M., O'Connor, K., & Wischnia, S. (2017). Knowing about racial stereotypes versus believing them. *Urban Education*, *52*(4), 491–524. doi:10.1177/0042085916672290

National Center for Education Statistics. (2006). *Economic impact of the nation's historically Black colleges and universities*. Washington, DC: U.S. Department of Education.

Newkirk, V. (2012). *New life for historically black colleges and universities: A 21st century perspective*. Jefferson, NC: McFarland & Co.

Nietfeld, J., Cao, L., & Osborne, J. (2006). The effect of distributed monitoring exercises and feedback on performance and monitoring accuracy. *Metacognition and Learning*, *2*(2), 159–179. doi:10.100710409-006-9595-6

Nietfeld, J., Shores, L., & Hoffman, K. (2014). Self-regulation and gender within a game- based learning environment. *Journal of Educational Psychology*, *106*(4), 961–973. doi:10.1037/a0037116

O'Cathain, A., Murphy, E., & Nicholl, J. (2010). Three techniques for integrating data in mixed methods studies. *BMJ (Clinical Research Ed.)*, *341*(7783), 1147–1150. doi:10.1136/bmj.c4587 PMID:20851841

Ong, M. (2005). Body projects of young women of color in physics: Intersections of gender, race, and science. *Social Problems*, *52*(4), 593–617. doi:10.1525p.2005.52.4.593

Ong, M., Wright, C., Espinosa, L. L., & Orfield, G. (2011). Inside the double bind: A synthesis of empirical research on undergraduate and graduate women of color in science, technology, engineering, and math. *Harvard Educational Review*, *81*(2), 172–208. doi:10.17763/haer.81.2.t022245n7x4752v2

Pajares, F. (1996). Self-efficacy beliefs in academic settings. *Review of Educational Research*, *66*(4), 543–578. doi:10.3102/00346543066004543

Palmer, R., Davis, R., & Maramba, D. (2010). Role of an HBCU in supporting academic success for underprepared Black males. *Negro Educational Review*, *61*(1-4), 85–124.

Parsons, E. C. (2014). Unpacking and critically synthesizing the literature of race and ethnicity in science education. In E. C. Parsons, N. G. Lederman, & S. K. Abell (Eds.), *Handbook of research on science education* (2nd ed., pp. 167–186). New York, NY: Taylor & Francis.

Parsons, E. C., & Mensah, F. M. (2010). Black feminist thought: The lived experiences of two Black female science educators. In K. Scantlebury, J. Butler Kahle, & S. N. Martin (Eds.), Re-visioning science education from feminist perspective (pp. 13–24). Rotterdam: Sense Publishers.

Purzer, S. (2011). The relationship between team discourse, self-efficacy, and individual achievement: A sequential mixed-methods study. *Journal of Engineering (Stevenage, England)*, *100*(4), 655.

Rao, V., & Flores, G. (2007). Why aren't there more African-American physicians?: A qualitative study and exploratory inquiry of African-American students' perspectives on careers in medicine. *Journal of the National Medical Association*, *99*(9), 986–933. PMID:17913107

Robnett, R., Chemers, M., & Zurbriggen, E. (2015). Longitudinal associations among undergraduates' research experience, self-efficacy, and identity. *Journal of Research in Science Teaching*, *52*(6), 847–867. doi:10.1002/tea.21221

Saldana, J. (2009). *The coding manual for qualitative researchers*. Los Angeles, CA: SAGE.

Scottham, K., Cooke, D., Sellers, R., & Ford, K. (2010). Integrating process with content in understanding African American racial identity development. *Self and Identity*, *9*(1), 19–40. doi:10.1080/15298860802505384

Sellers, R., Smith, M., Shelton, J. N., Rowley, S., & Chavous, T. (1998). Multidimensional model of racial identity: A reconceptual African American racial identity. *Personality and Social Psychology Review*, *2*(1), 18–39. doi:10.120715327957pspr0201_2 PMID:15647149

Seymour, E., & Hewitt, N. (1997). *Talking about leaving*. Boulder, CO: Westview Press.

Seymour, E., Hunter, A., & Larson, S. (2004). Establishing the benefits of research experiences for undergraduates in the sciences: First findings from a three-year study. *Science Education*, *88*(4), 493–534. doi:10.1002ce.10131

Shenton, A. (2004). Strategies for ensuring trustworthiness in qualitative research projects. *Education for Information*, *22*(2), 63–75. doi:10.3233/EFI-2004-22201

Shorette, C. R., & Palmer, R. (2015). Historically Black colleges and universities (HBCUs): Critical facilitators of non-cognitive skills for Black males. *The Western Journal of Black Studies*, *39*(1), 18–28.

Solórzano, D. G., & Yosso, T. J. (2002). Critical race methodology: Counter-storytelling as an analytical framework for education research. *Qualitative Inquiry*, *8*(1), 23–44. doi:10.1177/107780040200800103

Tabachnick, B. G., & Fidell, L. S. (2013). *Using Multivariate Statistics* (6th ed.). Boston: Allyn and Bacon.

Thomas, A., Hacker, J., & Hoxha, D. (2011). Gendered racial identity of black young women. *Sex Roles*, *64*(7-8), 530–542. doi:10.100711199-011-9939-y

Van Camp, D., Barden, J., Sloan, L., & Clarke, R. (2009). Choosing an HBCU: An opportunity to pursue racial self-development. *The Journal of Negro Education*, *78*(4), 457–468.

Williams, M., & George-Jackson, C. (2014). Using and doing science: Gender, self- efficacy, and science identity of undergraduate students in STEM. *Journal of Women and Minorities in Science and Engineering*, *20*(2), 99–126. doi:10.1615/JWomenMinorScienEng.2014004477

Zeldin, A., & Pajares, F. (2000). Against the odds: Self-efficacy beliefs of women in mathematical, scientific, and technological careers. *American Educational Research Journal*, *37*(1), 215–246. doi:10.3102/00028312037001215

Chapter 2
An Assessment of School Segregation and Its Impact on STEM Education:
New York City School Segregation

Tyresa Rene Jackson
https://orcid.org/0000-0002-4561-8766
Teachers College, Columbia University, USA

ABSTRACT

The focus of this chapter is to understand the intersection of school segregation and African American girls' performance in science, technology, engineering, and mathematics (STEM) education. This chapter will begin with an assessment of school segregation in New York City. The next section of the chapter will assess how school segregation influences the performance of African American female students in STEM education. The final section of this chapter will provide recommendations on how to increase the number of African American girls in STEM education.

INTRODUCTION

Improving science, technology, engineering, and mathematics education is a priority, especially among female students of color (White House Initiative on Educational Excellence for African Americans, 2016). By 2022, the American workforce will need one million additional STEM professions to meet workforce needs. The 2013 Women, Minorities, and Persons with Disabilities in the Science and Engineering Report by the National Science Foundation discovered that African American and

DOI: 10.4018/978-1-7998-4858-5.ch002

Latino participation in science and engineering occupations are low (Paula Johnson, 2016). To that point, in 2017, President Barack Obama launched the Educate to Innovate campaign to enhance the United States as a global leader in science and mathematics education (Blackburn, 2017). One of the three arching priorities of this campaign is to expand STEM education and career opportunities for underrepresented groups, including women and minorities. Studies show that developing an identity linked to STEM from a young age (Bieri Buschor, 2014), having a supportive family, access to quality advising, and exposure to gender-inclusive video games. Starting in elementary school, educators and school administrators alike must find innovative ways to develop, support, and propel African American girls into STEM education and careers. Women and minorities comprise 70% of college students but only attain 45% of STEM degrees (Office of Science and Technology Policy, 2016).

Education has long been an essential tool used to liberate and enlighten African American girls and women. More than sixty years after *Brown v Education* of Topeka which stated it was unconstitutional to segregate schools (Nelson, 2018), yet schools are now more segregated and as a result African American girls are greatly impacted and studies show, African American girls are six times more likely to be disciplined in school than boys and girls of any other race (Williams, Ocen, & Nanda, 2015). Schools segregated by race and socioeconomic status correlates to academic performance.

THE HISTORY OF SCHOOL SEGREGATION IN NEW YORK CITY

In 1958, a group of African-American mothers entitled, "The Little Rock Nine of Harlem," (Orfield, 2014, pg. 19) began a boycott to keep their children out of three Harlem junior high schools due to the schools' poor learning conditions (Back, 2003). As a result of defying the education compulsory law, this case landed in the Domestic Relations Court where a judge concurred with the opinion of the parents. Judge Polier's landmark decision was four years after the landmark *Brown v. Board of Education* case, and he charged the New York City Board of Education with offering an inferior education to African-American students. Following in the footsteps of the court, the Commission on Integration created a proposal where students from over-utilized schools in East Harlem and Bedford Stuyvesant, Brooklyn would have the opportunity to attend under-utilized schools in adjacent neighborhoods in Glendale, Queens and Yorkville on the East Side of Manhattan. In response to this matter, the Board of Education created limited open enrollment of junior high schools in 1960; this included 21 "sending" junior high schools with crowded classrooms and high concentrations of minorities to 28 "receiving" (Back, 2003) schools with space for

3,000 more students. The next year, this initiative expanded to elementary schools, but less than 700 students applied.

The movement to desegregate the public-school system across the United States was a major goal of the Civil Rights Movement (Library of Congress, n.d.). Since the Fair Housing Act of 1968, New York City remains sharply segregated in areas of housing and education (Walker, 2018). A report conducted by City Councilmember—Brad Lander, in partnership with Policy and Budget Director Annie Levers and several other City council members found New York's public transportation and infrastructure, paired with housing discrimination and other key issues have stagnated the City's integration (Mader, 2016). Neighborhoods separated by income often have school districts that are segregated, resulting in distinct differences in performance outcomes.

Allowing students from lower-income communities to attend school in more affluent communities can remove barriers to access and provide a platform to engage in challenging curriculum in science, mathematics, and technology. Schools that serve a disproportionate number of students of color offer fewer advanced math and science courses such as, calculus and physics (Sawchuk, 2018). Algebra 1 is considered a gateway course, and many Caucasian and Asian students are enrolled by the eighth grade. However, African-American and Native American students were disproportionately more likely to take Algebra 1 in their junior and senior year of high school, making it difficult for them to take more challenging mathematics courses during high school (Sawchuk, 2018). Taking science and mathematics is indicative of academic success during K-12 and beyond. According to Sawchuk (2018), students who take more mathematics and science courses are often better prepared for college.

School desegregation efforts throughout New York City have placed a focus on students who receive free or reduced lunch, a widely accepted measure that connects to poverty which has been correlated to race (Hu, 2018). Notably, the District 2 parent council points to research that shows students from low-income families are more likely to experience high levels of absenteeism (Veiga, 2017) due to factors like homelessness, which results in tardiness and poor attendance, placing these students at a distinct disadvantage. Segregation within the New York City public school system persists today in part due to de facto housing patterns that have replaced more explicit segregation policies of the past (Chen, 2018). According to Acevedo-Garcia and McArdle (2017), New York City public schools that are segregated by race and socio-economic level have distinct differences in student performance and outcomes. Schools facing high concentrations of poverty can improve achievement outcomes by having equitable access to resources, highly qualified teachers, and private and public funding (McArdle & Acevedo-Garcia, 2017).

Although schools are no longer segregated by race based on policies and laws, disparities continue to persist. Schools with a high proportion of students who receive free or reduced lunch are often less likely to have access to rigorous courses in the sciences and mathematics. Change the Equation (2015) examined the 2015 Nation's Report Card by assessing the gaps between students who attended the lowest-poverty schools (where no more than twenty-five percent of students qualify for no-cost or a reduced lunch) and the highest poverty schools (where at-least seventy-five percent of students qualify for lunch at no cost or a reduced price). According to the report, eighth-grade students who attended some of the nation's poorest schools have much less access to science labs and materials (Change the Equation, 2015). Further, the study also revealed eighth-graders who attended a middle-school where seventy-five percent or more of students are in poverty only forty-nine percent of teachers will have all or most the resources they need to teach science. On the other hand, schools with twenty-five percent or less of students in poverty will have seventy-five percent of teachers with all or most resources to teach science (Change the Equation, 2015).

HOW SEGREGATION INFLUENCES STEM PERFORMANCE

One of the root causes associated with school segregation is race and economic status (Vivian, 2016). The location of a school affects the amount of funding granted for technology and teachers' salaries. Reardon (2015) noted that neighborhood conditions appear to affect children's cognitive development and long-term educational outcomes. Neighborhood segregation is a factor that correlates to school segregation and may lead to achievement gaps if children are systematically placed in lower-and-higher income neighborhoods.

A 1966 Coleman Report, found segregated schools are less likely than affluent schools to offer Advanced Placement (AP) and honors courses and have been proven to provide less challenging curricula than schools in more affluent communities. Overall, many schools with high concentrations of low-income Black and Latino students are more likely to have inexperienced or unqualified teachers, fewer demanding college preparatory courses, more remedial courses, and higher teacher turnover (Lee, 2004). Goldberg (2008) highlighted that teachers become the most effective within the first three to five years of teaching. To that point, many experienced teachers are not equally distributed throughout low-income school districts and are more likely to gravitate toward schools with lower concentrations of poverty, minority populations, and high levels of student achievement (Boyd, Grossman, Lankford, Loeb, & Wyckoff, 2009). Many teachers who perform better on general knowledge certification examinations are more likely to have low-performing students, resulting in high turnover rates within lower-performing schools. As teacher turnover rate persists, this creates the challenge of building a school with a strong core of highly experienced teaching staff.

Performance outcomes in the classroom have continued to differ among races, as demonstrated in the 2011 National Assessment of Educational Progress which found 17% of African-American fourth grade students were proficient in mathematics and by eighth grade the number of students was lowered to only 15% of African-American students being proficient in mathematics (National Center for Education Statistics, 2012). The 2015-16 Civil Rights Data Collection Survey of Public Schools found that schools that had a majority of African-American or Latino enrollment were less likely to offer science and mathematics courses except algebra 1 and only thirty-eight percent of predominantly minority schools offered calculus (EdSource, 2018). Decades of educational research demonstrates that classrooms with mixed performance students can improve educational outcomes and opportunities (Hu, 2018). A study has found that parents who travel from their zone school districts traveled to schools that held the two following key features: (1) fewer poor students and (2) higher test scores. Many communities welcome the chance to pursue an education in distant communities but would prefer to have strong neighborhood schools (Fruchter, Mokhtar, Arvidsson, & Beam, 2015). A concurrent study at the end of court-ordered segregation in Charlotte-Mecklenburg, North Carolina found that minority students returning to a school closer to their residences had lower high school test scores and lower high school completion rates (Billings S. B., 2014). The School Composition and Black-White Achievement Gap Report found among eighth grade students who attended schools with more than sixty-percent of an African American population scored lower on national math tests (Barshay, 2015).

The system of tracking is a critical aspect of a students' learning experience because it affects the classes a student takes, the curriculum students are taught, the peer's students learn with, and who students are taught by. Tracking is a practice of grouping students together based on academic achievement (Loveless, 2009). Two studies of American students were conducted to compare the achievement of tracked and de-tracked students in the classroom and found many lower-income students often attend schools that are not tracked. DeSena and Ansalone (2009) noted once an African American student is identified as low performing, they are tracked from primary through secondary education, and placed in lower-level courses (DeSena & Ansalone, 2009; "Teaching Inequity", 1989). Additionally, social science has found internalizing gender stereotypes of being insufficient, leads to low performance in STEM courses (Girls Scouts of the USA/Girl Scout Research Institute, 2012). Furthermore, a report published by the University of California, Berkeley stated that diversity in science invigorates problem-solving skills and diversity helps to balance bias (The Scientific Community: Diversity Makes a Difference., n.d.).

Many public-school institutions in urban settings do not uniquely face challenges in the classroom, however, unlike schools in rural and suburban settings, schools in urban context are often in highly concentrated areas. Because of this, urban public education systems face a larger population of students who come from poverty, higher numbers of immigrant families, and frequent rates of student mobility (Kincheloe, 2010). Matters facing urban schools encompass both structural and cultural challenges. Thus, as we look at strategies and policies that are being instituted to remove educational barriers, and equalize opportunities for all children, there continues to be education policies and practices that have yet to find a cure to equalize opportunities for all.

Although the achievement gap among underrepresented groups - AfricanAmericans, Latinx, and Native Americans has decreased, underrepresentation continues to persist in the sciences and mathematics, engineering, and computer sciences (Archer, DeWitt, Osborne, & Dillon, 2010). In understanding this, the intersection of race and gender plays an important role in the identity one takes on within education, and beyond. Social scientists have found internalizing gender stereotypes of being insufficient, leads to low performance in STEM courses (Girls Scouts of the USA/Girl Scout Research Institute, 2012). As such, girls of color, particularly in urban learning environments develop negative ideas about their intellectual capabilities, resulting in a lack of interest in STEM. This, in turn, results in girls of color gravitating towards subjects that are viewed as more socially and culturally acceptable, which influences the kind of career one will pursue later in life. STEM is a field that encompasses tomorrow's future doctors, scientists, and engineers, with this in mind, many students shy away from the challenges of this field for an array of reasons—lack of education, gender stereotypes, and preconceived ideas.

AFRICAN AMERICAN GIRLS IN STEM EDUCATION

According to American College Testing (ACT) report—The Condition of STEM 2016, reported 30,057 African American students were tested in mathematics, science, and STEM, of those students for who were tested, twenty-five percent met ACT college math readiness standards, twenty-two percent met ACT science college readiness standards. In 2014 November, the Council and Women and Girls released a report entitled "Women and Girls of Color: Addressing Challenges and Expanding Opportunities", this report was created to assess challenges and barriers facing women and girls of color (Hansot, 1990). STEM skills have an increasingly important role in the American economy. Early on, both girls and boys show similar rates of interest in science, but the STEM achievement gap begins to widen starting in middle school and beyond (Williams H. A., 2005).

Disparities between boys and girls pursuing STEM-related courses are evident starting in primary school and are due to an array of factors, including societal, familial, and cultural influences. Although access to education for women and girls has improved globally, disparities in access to a basic education still persist, thereby influencing an additional disparity—the gender gap in STEM education. To demonstrate, a study conducted in the United Kingdom, found at ages ten to eleven both boys and girls equally engaged in STEM education (75% of boys and 72% of girls), and reported learning interesting things in science. Later, at the age of eighteen, these numbers changed to 33% of boys and 19% of girls learning interesting things in science (UNESCO, 2017). The importance of increasing women of color in STEM education and careers helps to support diverse scholarship and the dissemination of knowledge. Diversity serves as a platform to develop unique specializations in science and medicine. For example, some scientists may enjoy studying the cognitive functions of behavior (University of California, 2018), while other scientists focus their research on the biology of cancer and other diseases prevalent within communities of color. STEM is a field that encompasses tomorrow's future doctors, scientists, and engineers, with this in mind, many students shy away from the challenges of STEM due to—lack of education, gender stereotypes, and preconceived ideas.

ACHIEVEMENT GAPS FOR AFRICAN AMERICAN GIRLS IN STEM EDUCATION

A key factor in student success within a school is cultural competence—socially and psychologically. Cultural competency is defined as: "As a characteristic of good government. It promotes the effective delivery of services to a comprehensive and inclusive public." (Gooden, pg. 3, Cultural Competency for Public Administrators). Overall, research shows, cultural competency has developed as an integral part of service delivery and organizational change (Carrizales, 2010). Furthermore, cultural competence recognizes the cultural context in which the public encounters (Carrizales, Journal of Public Affairs, 16(4) 2017). Cultural differences among communities have led some governments to adapt efforts for increasing their cultural competency (Benavides & Hernandez, 2007). With increasingly changing demographics in the United States, public administration research has emphasized the need to explore cultural competency initiatives (Carrizales, 2010). White and Rice (2005) suggest the changing demographics of the country will require public organizations "to develop more inclusive work cultures that have an understanding

Figure 1. Percentage of Student Enrollment in STEM
(Source: (AAUW, 2010)

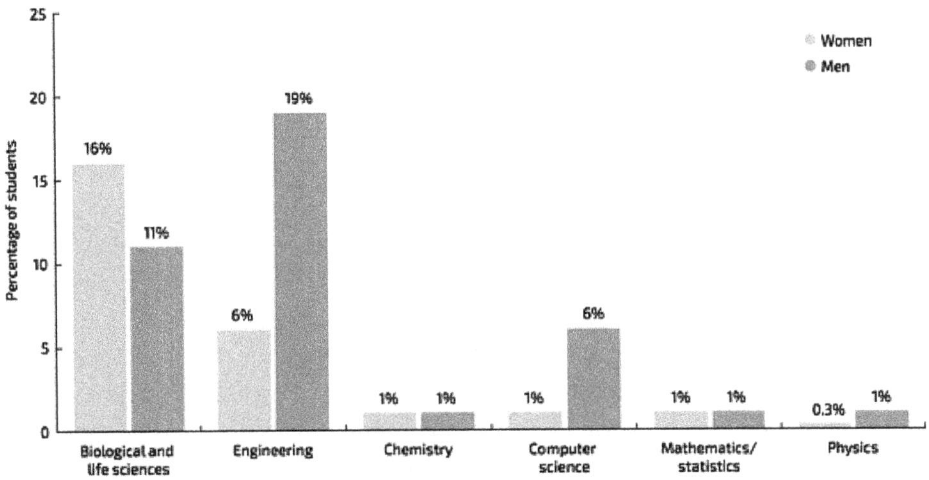

of the many ways people are different from one another and/or different from the organization." (p.3). Cultural competency initiatives within the public-school system will allow for increased effectiveness of future systems. Cultural competence reflects specific actions or policies within an organization that enable it to more effectively serve a culturally diverse population(s) (Carrizales, 2010). Moreover, Carrizales (2010) asserts the value of a "culturally competent professional", where students must become familiar with understanding different cultures and people.

The integration of gender and cultural competent curriculum into the public-school system enables educators and students alike to remove the veil of stigma and to understand the diversity of culture. Importantly, issues of institutional racism and discrimination remain an undercurrent issue within the United States. Recent polling about the trust between the government and minority groups demonstrates a large racial divide (Alkadry, Blessett, & Patterson, 2015). On balance, cultural competency initiatives within public school curriculum transcends into increased effectiveness in classroom participation and examination. According to the National Center for Cultural Competence (NCCC), cultural competency is defined as "having the knowledge, skills, and values to work effectively with diverse populations and to adapt institutional policies and professional practices to meet the unique needs of client populations." Thus, it is often an inherent belief that public school systems are neutral providers of goods and services within the education system, however, preconceived ideas of gender and race often influence service delivery.

To that point, the RTI Action Network (RTI Action Network, n.d.), discussed the importance of culture in the classroom with regard to cultural dissonance.

Practitioners use the family and community (i.e., poverty, and limited access to social and cultural capital) as the reason why underperforming low-income and racial/ethnic minority students struggle academically, therefore, cultural dissonance can be profoundly impactful, however, to the school experiences of urban students to assist with their overall learning development (Ahram, Stembridge, Fergus, & Noguera, 2018). Many times, it affects the interaction students and teachers have with one another; additionally, it contributes to misconceptions about students' motivations. Importantly, culture shapes and colors the expectations for achievement and sends important messages to students about how much (or little) their culture is valued by the school and larger society.

In addition to the aforementioned matters mentioned, the problem of inexperienced teaching staff is a pressing problem that many inner-city urban schools face. Many schools with high concentrations of low-income Black and Latino students are more likely to have inexperienced or unqualified teachers, fewer demanding college preparatory courses, more remedial courses, and higher teacher turnover (Lee, 2004). Researchers have found, teacher experience has an effect on student performance, and Goldberg (2008), highlighted studies that demonstrate teachers become the most effective the first three to five years of teaching. Yet, experienced teachers are not equally distributed throughout low-income schools, and studies reveal teachers are more likely to gravitate toward schools with low concentrations of poverty, low minority populations, and high levels of student achievement (Boyd, Grossman, Lankford, Loeb, & Wyckoff, 2009). Teachers who perform better on general knowledge certification examinations are significantly more likely to have low-performing students, in turn, leading to high turnover rates in low-performing schools. As the teacher turnover rate continues to persist in schools with a high number of low-performing students, this creates the challenge of building a school with a strong core of highly experienced teaching staff. Another important factor that is influential in the structure of urban education is data management systems. According to MDRC (2002), many urban schools have ineffective and underutilized data systems, making it difficult for school officials to monitor and track student performance and progress.

Studies show students within the inner city are more likely to absorb negative stereotypes about their own abilities—something that is particularly true when they are in segregated schools (Harvard Graduate School of Education, 2014). Nevertheless, students within the inner-city are likely to be surrounded by adults with low levels of educational attainment and limited professional prospects—a social context that has a very deep impact on a students' success within the realm of education (Schneider, 2017). Additionally, compared to their more-affluent peers, children's low-socioeconomic backgrounds read less frequently and are exposed to less complex language at home, which inhibits their cognitive abilities (Ayoub, et al., 2009) and their scores tend to be lower.

Low student achievement is often demonstrated on standardized state examinations, the Scholastic Aptitude Test (SAT), a small percentage of students taking Advanced Placement courses (AP classes) in high school, and reflected in a high number of students dropping out of high school. Many students within these schools express a desire to do well, yet barriers both in and out of the classroom impede on a student's progress, and often negative stereotypes that educators hold affect the interaction and expectation of students. The lack of instructional coherence often occurs as urban schools become bombarded with new instructional initiatives, and leads to fragmentation in curriculum implementation (Ahram, 2018). Moreover, since there is such a diverse student population within the inner-city, many urban school settings require diverse initiatives, nevertheless the approach for which the educators are using may further hinder curriculum implementation. Researchers recommend the curriculum is tailored to the broad needs of students within each school district and tailored to already existing systems.

The importance of providing students with positive experiences early on in their academic careers has lifetime benefits. A longitudinal study found having a strong elementary school experience can have a very positive effect on a student's learning and life trajectory, including reduced teenage pregnancy, higher levels of college attendance, and higher job earnings (Kuesera & Orfield, 2014). Science, Technology, Engineering, and Mathematics (STEM) is an important field that can lead to the development of future scientists, mathematicians, and engineering professionals who, in turn, cure diseases that are prevalent within low-income communities in the United States and throughout the world. Therefore, equalizing educational opportunities and allowing students from low-income and/or low-performing schools to integrate into schools that perform better academically, can lead to diversity in the field of science and engineering professions. High schools with the highest population of low-income students offer, on average, half as many AP math and science courses as schools with wealthier students (The Leadership Conference Education Fund, 2015). Often, the challenges of comprehension in the classroom begin early on, less than half of poor children and fifty-nine percent of children just above the poverty line are ready to start kindergarten (The Leadership Conference Education Fund, 2015).

Starting in elementary school, students attending minority and low-income are afforded fewer opportunities and instead of being taught by experienced teachers with strong backgrounds in reading instruction, math, and science students attend schools with higher concentrations of first-year teachers (3 to 4 percent). English learners also attend these schools at slightly higher rates (3 percent) than non-English learners (2 percent). Furthermore, students of color are also more likely to have teachers who aren't fully certified (U.S. Department of Education Office for Civil Rights, 2014). The shortcomings of many of these students' faces are demonstrated through standardized test scores. On the 2013 National Assessment of Educational Progress, higher-income Black and Hispanic students in fourth grade scored about the same or below low-income White and Asian students, using the federal free and reduced-price lunch program as a proxy for income (National Assessment of Educational Progress, 2013).

As we see these students reach middle school, circumstances don't differ much in performance. The National Science Foundation found that 36 percent of middle school science teachers and about 30 percent of middle school math teachers didn't have enough training in their subjects (The Leadership Conference Education Fund, 2015). The problem is that algebra is a gateway course to higher-level math in high school and beyond. Taking algebra in eighth grade puts students on track to reach calculus by their senior year in high school. Also, a third of all African-American students attend high schools that do not even offer calculus classes, compared with nineteen-percent of Caucasian students and thirteen-percent of Asian American students (U.S. Department of Education, 2014).

"'All of that starts in eighth grade, yet nationwide, Black and Hispanic students are much less likely to take or be placed in eighth-grade algebra. That opportunity gap is a big part of the reason why we see so few African-American and Hispanics youth in STEM majors in college" (Riegle-Crumb, 2014)

Due to No Child Left Behind, the public now has access to the records of students in K-12 academic performance. The data collected and reported, mostly consist of student standardized test scores. Today, the data that is collected is used—by policy leaders, parents, and the general public—as a measure of school quality, often in the total absence of other information. Research suggests test scores don't truly measure school quality (Schneider, 2017). School quality is only one factor among many that influences how a student performs—neighborhoods, family structures, and peer influence, each play a vital role in how a student performs within the classroom. To that end, discrepancies in test scores reveal a lot about families and neighborhoods than they do about the work being done in schools. For instance, a study found that the influence of family and neighborhood factors account for roughly sixty-percent of the variance in student test scores, and teachers, by contrast, account for only about ten-percent. In all, many students from the suburbs tend to enter school with the early literacy and numeracy skills necessary to learn the prescribed curriculum. Equally important, many students from well-resourced school districts learn school from role models at home and in the community (National Research Council Institute of Medicine, 2004). Many students from suburban and well-resourced communities hear positive school messages all the time and quickly develop the sense that doing well in school matters. When it comes time to take tests, such students tend to score quite well, and their schools play a vital role in the success of their students.

RECOMMENDATIONS ADDRESSING THE ACHIEVEMENT GAPS FOR AFRICAN AMERICAN GIRLS IN STEM EDUCATION

Overall, the benefits of supporting the implementation of a desegregation plan would ensure that admissions criteria, process, and mechanisms are transparent and easy to navigate, which can result in students from a wide array of demographics having equal access to education. Students placed in their zone school districts, in turn, can result in removing barriers to access to programs that support their development in math and science. Therefore, written below, are recommendations for ways to support the New York City Board of Education implementation of a desegregation plan and ways to schools that are not performing highly.

Streamline Mechanisms for Diversity Initiatives

New, online, mobile-ready applications for middle and high schools will help families without resources to navigate the current complex paper processes. Starting in elementary school, school officials and faculty should host local workshops to provide parents with the knowledge and tools to learn about open-choice schools and the benefits of students integrating into more diverse school districts.

Expand Diversity in the Admissions Process

Twenty-one elementary, middle and high schools have created Diversity in Admissions targets for special student groups – including students eligible for free or reduced-price lunch, English Language Learners – and students whose families were impacted by incarceration (New York City Department of Education). The Department of Education has developed holistic diversity plans that build upon the changes to their admissions programs through the Progressive Redesign Opportunity Schools for Excellence (PROSE) program.

Support Students in Temporary Housing

The Department of Education has partnered with the Department of Homeless Services and Human Resources Administration to create and implement a comprehensive strategy to increase access to Pre-K to 12 admissions processes for families in shelters (New York City Department of Education). This admission process supported over three-hundred shelter-based staff to inform patrons of shelter-based housing, with children, about educational services. Many Children within shelter housing are granted the opportunity to learn about resources in the community and in schools within the community. Therefore, working in conjunction with the Department of Homeless Services will allow students in temporary/shelter housing to have confined education and learning opportunities.

Create a Diversity Initiative When Opening New, High-Quality Schools and Programs

With the input and support from the School Diversity Advisory Group and communities, the Department of Education is intentionally building and maintaining diverse school communities by engaging with local stakeholders to the needs of students and the community, opening new unscreened schools, with a particular focus on schools with a campus. With this in mind, the New York City Department of Education can expand selection criteria when selecting students to enter different school districts.

Ensure Diversity and Equity is a Factor when Re-zoning and Creating Districts

Since there are constraints with regard to housing patterns and travel, when going through a rezoning process, the Department of Education should create equitable and diverse school zones a priority. This will create even more inclusive, collaborative planning processes for all our re-zonings. As demonstrated in the research, school funding is often correlated to the neighborhood location of the school. Therefore, when reconfiguring school districts and re-zoning, taking the placement of school districts into consideration can ensure more qualified teachers, funding, and after-school resources are afforded to school districts that have students from low-socioeconomic backgrounds.

Target Funding, Including Magnet Funding, to Foster Diversity

The Department of Education, over the years, has created and maintained magnet schools. As a result, magnet schools over a wide range of distinct programs have served students from many diverse backgrounds. Furthermore, under the federal Magnet Schools Assistance Program for 2015-16, the Department of Education was awarded supplemental funds to maintain seven magnet elementary schools in Districts thirteen, fifteen, and twenty-eight. Going forward, the DOE has decided to make an effort to target funding to foster diversity, this will enable school officials to not only support diversity initiatives in under-resourced but also, this will allow officials to address matters within low-performing schools. For example, low-performing schools can pay to send more teachers to training or go for further certifications/education.

CONCLUSION

In many ways, classroom diversity goes hand-in-hand with school quality. As the New York City Department of Education increases diversity in partnership with the School Diversity Advisory Group and districts, it is important to develop and expand resources to all students throughout the city. The New York City diversity plan is part of our Equity and Excellence for All agenda, along with the special diversity programs to promote AP for All and Computer Science for All (New York City Department of Education). In 2015, the Equity and Excellence for All agenda was announced by Mayor de Blasio and Chancellor Fariña, this initiative works to create an earlier foundation for student success, expand rigorous coursework to every school and provide more socio-emotional supports and college-awareness and planning supports for all students. Overall, desegregating schools can serve to support students in expanding their knowledge in STEM and other core competencies, resulting in future diversity, public health matters, science, and technology.

If we are to increase the likelihood of African-American students thriving in STEM or other STEM-related fields we must cultivate an educational environment that increases intellectual aptitude by incorporating calculus, chemistry, physics as part of the mandatory curriculum starting in primary education With the support of family, teachers, and positive adults, low-income inner-city urban students can dismember negative stereotypes and cultivate a generation scientist and mathematicians. Teachers and faculty alike, starting from elementary through post-secondary must provide additional supports (e.gSTEM afterschool programs and culturally competent class material); further, recruiting more women teachers of diverse cultures who are educated in a STEM-related field, in turn, removes the stigma students' inner-city urban schools are not smart enough. To conclude, parents and caretakers alike are encouraged to place children in STEM after-school programs and summer camps, which increases intellectual abilities-critical thinking, mathematics skills, and reading and to attend local community boards to allow their children to participate in opportunities beyond what is offered in their neighbor school district.

REFERENCES

AAUW. (2010). *Why so few? Women in science, technology, engineering, and mathematics.* AAUW.

Ahram, R., Stembridge, A., Fergus, E., & Noguera, P. (2018). *Framing Urban School Challenges: The Problems to Examine When Implementing Response to Intervention.* Retrieved from RTI Action Network: http://www.rtinetwork.org/learn/diversity/urban-school-challenges

Alabama State Archives: Slaves and Free Persons of Color. (1833). Retrieved July 2018, from Alabama State Archives: http://www.archives.state.al.us/cornerstone/slavecode1833/page07.html

Alkadry, M. K., Blessett, B., & Patterson, V. L. (2015). Public Administration, Diversity, and the Ethic of Getting Things Done. *Administration & Society*, 1191–1218.

Aptheker, H. (1974). The Education of Black People: Ten Critiques, 1906-1960. *The Journal of Negro Education*, *43*(1), 135–138. doi:10.2307/2966953

Archer, L., DeWitt, J., Osborne, J., Dillon, J., Willis, B., & Wong, B. (2010, July). "Doing" Science Versus "Being" a Scientist: Examining 10/11-Year-Old School Children's Constructions of Science Through the Lens of Identity. *Science in Education*, *94*(4), 617–639. doi:10.1002ce.20399

Ayoub, C., O' Connor, E., Rappolt-Schlictmann, G., Vallotton, C., Raikes, H., & Chazan-Cohen, R. (2009). Cognitive Skill Performance Among Young Children Living in Poverty: Risk, Change, and the Promotive Effects of Early Head Start. *Early Childhood Research Quarterly*, *24*(3), 289–305. doi:10.1016/j.ecresq.2009.04.001

Back, A. (2003). Exposing the "Whole Segregation" Myth: the Harlem Nine and New York City's public schools desegregation battles. In K. Wood (Ed.), *"Freedom North," Black freedom struggles outside the south 1940-1980* (pp. 65–91). Gordonsville, VA: Palgrave Macmillan. doi:10.1007/978-1-4039-8250-6_4

Barshay, J. (2015, September 18). *Black Math Scores Lag the Most in Segregated Schools.* Retrieved from The Hechinger Report: https://hechingerreport.org/black-math-scores-lag-the-most-in-segregated-schools/

Bell, A., Chetty, R., Jaravel, X., Petkova, N., & Van Reenen, J. (2018). *Who Becomes an Inventor in America? The Importance of Exposure to Innovation.* Harvard University.

Benavides, A. D., & Hernandez, J. C. (2007, July). Serving Diverse Communities-Cultural Competency. *Public Management*, 14–18.

Bieri Buschor, C. S., Berweger, S., Keck Frei, A., & Kappler, C. (2014). Majoring in STEM - What Accounts for Women's Decision Making? A Mixed Methods Study. *The Journal of Educational Research, 3*(107), 167–176. doi:10.1080/00220671.2013.788989

Billings, S. B., Deming, D. J., & Rockoff, J. (2014). School Segregation, Education Attainment, and Crime: Evidence from the End of Busing in Charlotte-Mecklenburg. *The Quarterly Journal of Economics, 129*(1), 435–476. doi:10.1093/qje/qjt026

Blackburn, H. (2017). The Status of Women in STEM in Higher Education: A Review of the Literature 2007–2017. *Science & Technology Libraries, 36*(3), 235–273. doi:10.1080/0194262X.2017.1371658

Boyd, D., Grossman, P., Lankford, H., Loeb, S., & Wyckoff, J. (2009). Teacher Preparation and Student Achievement. *Educational Evaluation and Policy Analysis, 31*(4), 416–440. doi:10.3102/0162373709353129

Carrizales, T. (2010). Exploring Cultural Competency Within the Public Affairs Curriculum. *Journal of Public Affairs Education, 16*(4), 593–606. doi:10.1080/15236803.2010.12001616

Change the Equation (2015). *Ending the Double Disadvantage.* CTEq analysis of U.S. Department of Education.

Chen, M. (2018). *New York's Separate and Unequal Schools.* Retrieved, from The Nation: https://www.thenation.com/article/new-yorks-separate-and-unequal-schools/

Citizens' Committee for Children of New York. (2016, March 31). *Understanding Diversity in NYC's Public Schools.* Retrieved from https://www.cccnewyork.org/blog/understanding-diversity-in-nycs-public-schools/

Darville, S. (2018, April 27). *A Chalkbeat Cheat Sheet: What's Going With Upper West Side Desegregation.* Retrieved from Chalkbeat: https://chalkbeat.org/posts/ny/2018/04/27/a-chalkbeat-cheat-sheet-whats-going-on-with upper-west-side-desegregation/

Demause, N. (2016, September 14). *Desegregating NYC's Apartheird Schools.* Retrieved from The Village Voice: https://www.villagevoice.com/2016/09/14/desegregating-nycs-apartheid-schools/

EdSource. (2018). *Latino, African-Americans Have Less Access to Math, Science Classes, New Data Show.* Retrieved from EdSource: https://edsource.org/2018/latino-african-americans-have-less-access-to-math-science-c asses-new-data-show/598083

Epperly, J. (2014). New York Schools Most Segregated in the Nation. University of California, Los Angeles. The Wakeman Agency.

Fruchter, N., Arvidsson, T., Mokhtar, C., & Beam, J. (2015). *Demographics and Performance in New York City's School Networks.* An Initial Inquiry, Brown University, Annenberg Institute for School Reform.

Gurria, A. (2018). *PISA 2015 Results in Focus.* OECD.

Harris, L. M. (2003). *In the Shadow of Slavery: African Americans in New York City 1626-1863.* Chicago, IL: The University of Chicago Press.

Harvard Graduate School of Education. (2014). *Claude Steele-Stereotype Threat.* Retrieved from Harvard School of Education: https://www.gse.harvard.edu/news/14/11/claude-steele-stereotype-threat

Hemphill, C., & Mader, C. (2016). Segregated Schools in Integrated Neighborhoods: The City's Schools and Even More Divided Than Our Housing. Center for New York City Affairs, The New School.

Hu, W. (2018, June 6). *In a Twist, Low Scores Would Earn Admission to Select Schools.* Retrieved from New York Times: https://www.nytimes.com/2018/06/06/nyregion/middle-school-admission-desegregation nyc.html

Hu, W. (2018, May 17). *'It Has to Start Somewhere': Grass-Roots Drive to Integrate New York Schools.* Retrieved September 15, 2018, from Nytimes.com: https://www.nytimes.com/2018/05/17/nyregion/it-has-to-start-somewhere-grass-roots-dr ve-to-itegrate-new-york-schools.html

Irons, P. (2004). *Jim Crow Schools.* Retrieved from American Teachers Federation: https://www.aft.org/periodical/american-educator/summer-2004/jm-crows-schools

Kincheloe, J. L. (2010). Why a book on urban education? In *19 Urban Questions: Teaching in the City; Forward Antonia Darden* (pp. 1–32). New York, New York: Peter Lang Publishing.

Kucsera, J., & Orfield, G. (2014). *New York State's extreme school segregation: Inaction, inequality, and a damaged future.* Los Angeles, CA: Civil Rights Project/Proyecto Derechos Civiles.

Kuesera, J., & Orfield, G. (2014). *New York State's Extreme School Segregation.* The Civil Rights Project.

Learned-Miller, C. (2016, October 14). *Cambridge Public Schools: Pioneers of Equitable Choice*. Retrieved from The Century Foundation: https://tcf.org/content/report/cambridge-public-schools/

Lee, C. (2004). Racial Segregation and Educational Outcomes in Metropolitan Boston. The Civil Rights Project, Harvard University.

Library of Congress. (n.d.). *Civil Rights History Project*. Retrieved from Library of Congress: https://www.loc.gov/collections/civil-rights-history-project/articles-and-essays/school-se regulation-and-integration/

Literacy as Freedom. (n.d.). Retrieved from Smithsonian American Art Museum: https://americanexperience.si.edu/wp-content/uploads/2014/09/Literacy-as-Freedom.pdf

Loveless, T. (2009). *Tracking and Detracking: High Achievers in Massachusetts Middle Schools*. Dayton, OH: Thomas B. Ford Foundation.

Lyons, P. (2002). *Jim Crow's School*. Retrieved from American Federation of Teachers https://www.aft.org/periodical/american-educator/summer-2004/jim-crows-schools

MAPP. (n.d.). Retrieved July 2018, from Mapping African American Past: https://maap.columbia.edu/place/9.html

McArdle, N., & Acevedo-Garcia, D. (2017). *Consequences of Segregation for Children's Opportunity and Wellbeing*. Harvard University, Joint Center for Housing Studies of Harvard University.

MDRC. (2002). *Foundations for success: Case studies of how urban school systems improve student achievement*. Washington, DC: Author.

National Assessment of Educational Progress. (2013). *4th grade Math Scores by Race, Ethnicity and Income*. Retrieved from http://nces.ed.gov/nationsreportcard/naep- data/report.aspx

National Center for Education Statistics. (2012). *The Condition of Education 2012*. https://nces.ed.gov/pubs2012/2012045.pdf

National Research Council Institute of Medicine. (2004). The Nature and Conditions of Engagement. In N. R. Medicine (Ed.), *Engaging Schools: Fostering High Schools Students' Motivation to Learn* (pp. 31–59). Washington, DC: The National Academies Press.

Nelson, R. (2018). *Still Separate, Still Unequal: The Role of Black Women and Girls in the Legacy of Brown v Board of Education*. Retrieved from AAUW: https://www.aauw.org/2018/05/17/still-separate-still-unequal/

New York City Department of Education. (n.d.). *Equity and Excellence for All*. Diversity in New York City's Public School.

NYC Department of Education. (2018, June 20). *Chancellor Carranza Announces District 3 Middle School Diversity Plan*. Retrieved from NYC Department of Education: https://www.schools.nyc.gov/about-us/news/announcements/contentdetails/2018/06/20chancellor-carranza-announces-district-3-middle-school-diversity-plan.

Office of Science and Technology Policy. (2016). *STEM Depiction Opportunities*. Retrieved from Office of Science and Technology Policy https://obamawhitehouse.archives.gov/ sites/default/files/microsites/ostp/imageofstemdepictiondoc_02102016_clean.pdf

Paula Johnson, M. (2016). *STEM Pathways for Girls of Color - A Review of the Literature*. Intercultural Development Research Association.

Ratner, G. (2007). Why the No Child Left Behind Act Needs to Be Restructured to Accomplish Its Goals and How to Do It. *University of the District of Columbia Law Review, 9*(1).

Riegle-Crumb, C., & Moore, C. (2014). The Gender Gap in High School Physics: Considering the Context of Local Communities. *Social Science Quarterly*, *95*(1), 253–268. doi:10.1111squ.12022 PMID:25605978

RTI Action Network. (n.d.). *Framing Urban School Challenges: The Problems to Examine When Implementing Response to Intervention.* Retrieved from RTI Action Network: http://www.rtinetwork.org/learn/diversity/urban-school-challenges

Sawchuk, S. (2018). *Students of Color Face Persistent Disparities in Access to Advanced STEM Courses*. Retrieved from Education Week: http://blogs.edweek.org/edweek/curriculum/2018/04/persistent_disparities_in_stem_coursetaking_crdc.html

Schenk, D. H. (2014). White Terrorism and Black Disarment During Reconstruction. *Gettysburg College Journal of the Civil War Era, 4*(4).

Schneider, J. (2017). *The Urban School Stigma*. Retrieved from The Atlantic: https://www.theatlantic.com/education/archive/2017/08/the-urban-school-stigma/53796 /

The Leadership Conference Education Fund. (2015). *Advancing Equity through More and Better STEM*. The Leadership Conference.

The Scientific Community: Diversity Makes a Difference. (n.d.). Retrieved from University of California, Berkeley. https://undsci.berkeley.edu/article/socialsideofscience_02

The United States Constitution. (n.d.). *The U.S. Constitution and Fascinating Facts About It* (T. L. Jordan, Ed.). Naperville, IL: Oak Hill Publishing Company.

The White House Council on Women and Girls. (2015). *Advancing equity for women and girls of color*. Washington, DC: Author.

Tyack, D. B., & Hansot, E. (1990). *Learning together: A history of coeducation in American schools*. New Haven: Yale University Press. doi:10.2307/j.ctt1ww3vbg

University of California. (2018). *The Scientific Community: Diversity Makes a Difference*. Retrieved from https://undsci.berkeley.edu/article/socialsideofscience_02

U.S. Department of Education. (2014). Expansive Survey of America's Public Schools Reveals Troubling Racial Disparities. U.S. Department of Education, Office for Civil Rights data collection.

U.S. Department of Education Office for Civil Rights. (2014). *Civil Rights Data Collection, Data Snap- shot: Teacher Equity*. Retrieved from http://www2. ed.gov/about/of ces/list/ocr/docs/crdc-teacher- equity-snapshot.pdf

Veiga, C. (2017, November 7). *Want to Make Middle School Admissions More Fair? Stop Looking at this Measure, Parents Say*. Retrieved from Chalkbeat: https://www.chalkbeat.org/posts/ny/2017/11/07/want-to-make-middle-school-admission -more-fair-stop-looking-at-this-measure-parents-say/

Vivian, L. (2017). The Influence of School Segregation on Students' Educational Achievement. *Perspectives*, *9*(1), 11. Available at https://scholars.unh.edu/perspectives/vol9/iss1/11

White House Initiative on Educational Excellence for African Americans. (2016). *Fact Sheet: Supporting Science, Technology, Engineering, and Mathematics (STEM) Success Among African American Students*. U.S. Department of Education.

Wilder, C. S. (2013). *Ebony and Ivy*. New York, NY: Bloomsbury Press.

Williams, H. A. (2005). *Self-taught: African American Education in Slavery and Freedom*. Chapel Hill, NC: The University of North Carolina Press.

Williams, K., Ocen, P., & Nanda, J. (2015). *Black Girls Matter: Pushed Out, Overpoliced, and Underprotected*. Columbia University, Center for Intersectionality and Social Policy Studies.

Chapter 3
Celebrating Transformations Through STEM Storytelling

Pamela M. Leggett-Robinson
PLR Consulting, USA

ABSTRACT

Storytelling describes the social and cultural activity of sharing stories. Every culture has its own stories or narratives, which are shared as a means of entertainment, education, cultural preservation, or instilling moral values. In the STEM culture, women of color tell their stories about overcoming barriers and obstacles. Their stories echo familiar themes (the lack of role models, gender bias, and/or cultural racism within the organization/institution) as well as shed light on decades of programs continuing to fall short in empowering and encouraging women of color, thus resulting in an exodus from the STEM pipeline. More disturbing is their stories are often silenced or filled with embellishments and improvisations by the institutions or organizations they faithfully served. The chapter reflects on a "typical" story about a woman of color in STEM higher education. It takes the reader on a journey from entry to departure from the "fairy tale pipeline" of higher education. It is a story of struggle, transformation, and love, within the culture of STEM.

INTRODUCTION

Despite renewed efforts of initiatives, policies, research activities, professional organizations focusing on women of color in STEM, and national funding within the last twenty years, women of color remain underrepresented in science, technology, engineering, and mathematics (STEM) disciplines (National Academy of Sciences 2011; National Science Foundation 2017). While much research has focused on ways

DOI: 10.4018/978-1-7998-4858-5.ch003

to recruit women into these disciplines, less work has explored the strategies women use to navigate these STEM contexts (Lezotte and Ferguson, 2017; Leggett-Robinson and Villa, 2019) or the lasting results of having to navigate the STEM landscape on their professional and personal lives. Even more disconcerting is that programs focused on increasing and retaining women of color in STEM fields continue to fall short, resulting in an exodus from traditional to non-traditional STEM professions.

Demographic Changes Promise An Even Larger Loss Of Talent

National demographic data show that a full 92 percent of U.S. population growth in the last decade came from people of color (Cárdenas and Treuhaft, 2013). By 2050, communities of color will be in the majority, representing a projected 53 percent of the U.S. population. Growth of the young Latino population will continue its sharp rise, with a projected 63 percent increase in public high school graduates between the years 2008-09 and 2020-21 (NCES, 2013). Contrast this with the small number of Latinas earning STEM bachelor's degrees: in 2017, 3.5 percent of STEM degrees overall and 2.1 and 1.7 percent of engineering and computer science degrees, respectively, went to Hispanic women (NCES, 2017).

Further, Black women earned a smaller percentage of bachelor's degrees than their hispanic counterparts across all STEM fields e.g., Latinas, 3.8% and Blacks, 2.9%. (NCES, 2017).

The STEM community has sought to address the issue of low representation for WOC using diverse approaches, including efforts that work toward mentoring, networking, building personal agency, and leadership development. This chapter utilizes the simplicity of storytelling (A STEM Autobiography) to address the familiar and formidable barriers faced (STEM belonging, microaggressions, gender bias, and cultural racism) by women of color in STEM fields and to celebrate their personal and professional transformations.

THE "LEAKY PIPELINE"

The STEM pipeline metaphor assumes a linear progression through higher education and STEM occupations. Studies continue to assert that there are **not enough** women of color (WOC) graduating with degrees in STEM, and thus **not enough** WOC entering the traditional STEM workforce. Recent research suggests the "leaky pipeline" of underrepresentation of WOC in STEM is not just a matter of increasing entry into the pipeline (Whittaker & Montgomery, 2014; Sethna, 2011) but also a matter of retention at every educational transition point and in career progression. The lack of WOC in STEM is not a K-12 problem, a higher education problem, a government or private industry problem. It is a societal problem rooted in systems of inequity. Wolf-Wendel and Ward (2016) state, "Instead of blaming the drop of water for leaking out of the pipe; it is more productive to figure out why the pipe is leaking in the first place".

Documented causes for the leaking pipeline are the structural barriers within the culture of STEM. These barriers consist of but are not limited to stereotype threat, gender bias, lack of role models, discrimination, tokenism, and self-presentation (Leggett-Robinson & Villa, 2019). Additionally, the intersectionality of race, gender, and the culture of STEM for WOC make it extremely difficult to pinpoint which of the aforementioned is cause for the lack of retention in STEM fields. The combination of intersectionality and STEM barriers compounded over time, ultimately lead to a lack of STEM belonging and feelings of isolation for WOC, thus resulting in "leaks" from the pipeline.

By examining the intersectionality of race and gender within the STEM culture, researchers begin to understand the distinctive marginalization experienced by WOC in the STEM workplace (Alfred, 2019; Ireland, 2018). These experiences shape the social identities and psychological processes of WOC. Addressing the structural barriers within the STEM culture is essential to understanding ***why*** the underrepresentation of WOC in STEM continues (Snyder, 2014; Kang, 2019) (Table 1).

NONLINEARITY AND WOC IN STEM

STEM careers can be generally classified into five groups: computer occupations, engineering occupations, life & physical science occupations, social science occupations, and mathematical occupations. The current pipeline assumes a linear progression to achieve success in any of the aforementioned STEM careers; however, intersectionality and structural STEM barriers make linear progression more difficult for WOC, especially in more male-dominated occupations of computer science, engineering, physics, and math.

A linear career path is one that constantly moves in one direction and has progression of job responsibilities within an industry; it follows a traditional trajectory of moving up in power or status over time. However, a linear career path is not always possible as sometimes people can become stuck in a certain position and not progress further.

For many WOC in STEM, their thoughts about career progression are straightforward: graduate from a university, take a job of sorts, and then start the endless climb to success. Unfortunately, once they enter the workforce, life events occur and they choose to deviate from the prescribed formula of success in STEM and follow a non-linear career path.

Table 1. Education and Career Progression "Leaks" in the STEM Pipeline for WOC

Career Progression	Causes of Leaks to the STEM Pipeline for WOC
Primary and Secondary Education	*-stereotype threat || stereotypes* *-bias (gender, race, ethnic)* *-lack of role models* *-lack of STEM understanding* *-family/cultural expectations* *-lack of STEM career options*
Post-Secondary Education	*-stereotype threat || stereotypes* *-bias (gender, race, ethnic)* *-discrimination* *-harassment* *-lack of role models* *-lack of belonging & isolation*
Early Career	*-flawed recruitment practices* *-discrimination* *-harassment* *-childbearing/rearing caregiving responsibilities* *-lack of role models* *-lack of mentors* *-lack of job security (presumed)* *-lack of flexible work arrangements* *-stereotype threat || stereotypes*
Mid Career	*-retraining or re-entry into workforce can be difficult* *-lack of mentors* *-lack of sponsors* *-career interruptions* *-lack of flexible work arrangements* *-adult caregiving responsibilities*
Senior Level Career	*-increased pressure to sit on boards and committees to meet diversity requirements (tokenism & burnout)* *-bias (gender, race, ethnic)* *-adult or spouse caregiving responsibilities* *-discrimination/sexism*

According to Hill et.al (2014), different types of life events, e.g. late start, caregiving, and health problems, may produce nonlinearity for WOC in STEM professions. These life events are classified as pauses where time was taken out of the linear trajectory of one's STEM career, diversions that forced or slowed career progress, or hijacked advancement altogether. In both academic and private industry STEM careers, the 'sacrifice of time' is 'the most rigorously necessary condition for the accumulation of symbolic capital known as a reputation for professional worthiness' (Bourdieu, 1988, pg. 96). While a linear STEM career can be measured objectively and celebrated by the STEM professional community, Bourdieu's assertion of 'professional worthiness' resonates negatively with WOC in STEM who must 'pause' their careers due to life events outside of their control. STEM careers are 'pretty unforgiving' of the time-out necessitated by parenting or caregiving and thus 'borde[r] on being discriminatory' (Ward & Wolf-Wendel, 2012).

According to research by O'Connor & Cech (2018), 28% of those women and WOC who become parents leave full-time STEM employment within seven years and almost 15% of mothers left the workforce entirely. The incompatibility between meeting the demands of a STEM workplace culture, feelings of professional unworthiness because of caregiving responsibilities, and unfulfilled career goals may account for the choice talented WOC in STEM make to pursue positions outside the academy, i.e. consulting or tech start-ups. These positions allow a blend of productive and meaningful work with a satisfying personal life, and encourage creativity, collaboration, and connection (Paludi, 2008). In making this choice, WOC create lives and careers in a progressive and often non-linear fashion due to the barriers within the STEM culture.

THE STEM CULTURE AND CLIMATE

According to Lawrence and Keleher (2004), institutional (organizational) racism in the U.S. is the normalization and legitimization of an array of dynamics – historical, cultural, institutional and interpersonal. These dynamics routinely advantage whites (male and female) while producing cumulative and chronic adverse outcomes for people of color. It is a system of hierarchy, privilege, and power for one racial group at the expense of an oppressed people. In STEM (higher education or workplace), this system of advantage takes the form of cultural values, organizational policies, and procedures, which contribute to benefits for white people and the exclusion of women of color. It can be seen or detected in processes, attitudes, and behaviors which amount to discrimination through unwitting prejudice, ignorance, thoughtlessness, and racist stereotyping which disadvantage women (especially) of minority ethnicity (Leggett-Robinson & Villa, 2019).

Institutional culture represents the collection of shared values and beliefs that is the blueprint that guides actions, which inevitably establish climate. *Institutional climate*, however, represents the practices and behaviors that determine the prevailing attitudes in the environment.

The alignment of culture and climate in STEM (higher education and workforce) have traditionally privileged norms that favor and are associated with White males (Johnson et.al 2011; Ko et.al 2015, Ong 2005). This privilege extends to structures that govern learning, mark progress in educational attainment, and afford opportunity in the workforce, each typically associated with STEM mainstream participation, persistence, and success (Cress and Sax, 1998). This alignment coupled with the social and interpersonal factors of STEM belonging, microaggressions, and gender bias, account for the disproportion of WOC in STEM persisting at the same rates as their white male or female counterparts (Brown, 2000; Carlone & Johnson, 2007; Ong et al., 2011; Valenzuela, 2006; Varma, 2002; Varma et al., 2006).

The alignment of culture and climate in STEM affect men, women, and WOC differently. To align the culture and climate of STEM such that all members have the capacity to thrive, there must be a "shift" in organizational thought leadership to include those salient cultural aspects that impact men, women, and WOC's careers. If not, the misalignment will continue to result in pipeline leaks for historically marginalized minorities (Allen-Ramdial & Campbell, 2014).

STEM Belonging

The human need to belong is among the most powerful motivators of social behavior (Baumeister & Leary 1995). Because this need is so ingrained, we constantly monitor our belonging status and naturally and subconsciously detect even very subtle indications (e.g., lack of eye contact) that our belonging is threatened (Wesselmann et.al, 2012; Kerr & Levine 2008; Lewis et.al 2016). When people perceive threats to their belonging (e.g., when they feel excluded, rejected, or isolated), their day to day functioning can be negatively impacted.

STEM Belonging and Identity in Higher Education

STEM belonging refers to the sense of being accepted, valued, included (legitimate members in their academic domain), and encouraged by others (Goodenow 1993, p. 80) within the STEM community. In addition to self-beliefs, STEM students' connections to their discipline community can enhance their academic engagement and subsequently impact their identification within the STEM community (Fleming, 1984; Good, 2012; Johnson, 2011, 2012).

This connection covers both a sense of belonging to an academic setting (an institution, a department, or subgroups within them) and a psychological sense of community, a broader connection to the discipline or field area (National Academies of Sciences, Engineering, and Medicine, 2016).

Self-perceptions are framed by personal and collective identities. Each student has many such identities - racial, ethnic, socioeconomic, sexual/gender, and family. These identities are framed by upbringing and experiences. Students' positive identification in STEM can enhance academic engagement and belongingness and prove to be a great source of engagement as well as a motivator for persistence in obtaining a STEM degree. More common however, the antithesis is true, especially for historically marginalized student groups (Leggett-Robinson, Davis, & Villa 2018). These students typically experience challenges such as isolation, invisibility, a sense of not belonging and disconnectedness from external social and cultural networks (Ong 2001; NRC 2012)

STEM Belonging for WOC

Belonging to valued social groups is a fundamental human need; a sense of inclusion is particularly important for WOC in STEM when the culture and climate in which they operate imply they might be unsuited, higher education and workforce (Leggett-Robinson, Davis, & Villa 2018). Feeling a sense of belonging and acceptance by others in the STEM community is crucial to the retention of WOC in STEM fields.

As mentioned, a well-known barrier faced by WOC in their STEM progression is the sense of not belonging and struggling with conflicting identities (Leggett-Robinson & Villa 2019). This barrier is perhaps most easily understood through the lens of "social identity threat," or the feeling of isolation in the STEM community because of one's gender and/or race (Steele et al 2002). Although social identity is a person's sense of who they are based on their group membership(s), it has been argued that identity is not a construct that remains fixed in the mind; rather, individuals progress through their own identity journey in response to the social context, i.e. STEM. WOC who successfully navigate through undergraduate and/or graduate STEM education and land a position in the STEM workplace have presumably traversed through the various stages of self-identity.

Research indicates that a sense of belonging, or lack thereof, in STEM has a significant impact on educational and professional success and persistence, especially for WOC (Rainey et al 2018). WOC consistently report less sense of belonging than men and white women (Good et al 2012; Johnson 2012; Smith et al 2013) and report feeling unsure of their ability to "fit in" (Smith et al 2013). Such feelings often cause WOC to experience imposter syndrome, undermine professional performance, and give rise to perceptions that people like them do not belong there. Belonging and identification are most often lumped together in definition however, when considering the experience of WOC in STEM it is possible to be STEM identified while simultaneously lacking in a sense of belonging.

Microaggressions

Sue et al (2007a) describes microaggressions as "brief and commonplace daily verbal, behavioral, or environmental indignities, that communicate hostile, derogatory, or negative racial slights and insults toward people of color" of which the perpetrators are often unaware. Although not a new form of racism, racial microaggressions represent a more subtle or aversive racism where the racist intentions are covert (DeAngelis 2009). Microaggressions can also be described as "subtle snubs or dismissive looks, gestures, and tones" (Sue et al 2007a). Within STEM, common microaggressions against WOC from colleagues or supervisors include subtle sexist or racist remarks (Kachchaf et al 2015; Parker 2013; Shehab et al 2007; S. W. Brown 2008). Attitudes and behaviors are considered 'micro' because of the 'quiet' form in which they occur compared to forms of overt racism. Despite their quiet nature, the impacts of racial microaggressions have large implications for the WOC that receive them (DeCuir-Gunby 2019).

Racism continues to be a pertinent issue for WOC in STEM and experiencing racial microaggressions is quite commonplace in the STEM landscape (Louis et al 2016). The racial microaggressions experienced by WOC while pursuing a STEM undergraduate degree continue into graduate school and then into the STEM workplace. WOC experience profound physical, mental, and emotional consequences as a result of microaggressions (Smith, Hung, & Franklin 2011; Wang, Leu, & Shoda 2011). These consequences create health related issues (e.g. high blood pressure) and increased feelings of isolation or problem avoidance. In an effort to cope, they must choose whether to ignore the microaggressions or respond. However, by responding, WOC risk being further marginalized and seen by colleagues as "too sensitive" or "angry and aggressive" (Yosso et al 2009) which have ramifications in relationships, collaborative work, intellectual respect, receiving critical resources, support, and advancement opportunities (Charleston, Adserias, Lang, & Jackson 2014). The latter is rarely seen as the best coping mechanism, thus the former is more frequently utilized but with byproducts of increasing stress, social withdrawal, and self-criticism.

In STEM fields, microaggressions can move beyond interpersonal interactions. *Institutional microaggressions* have been defined as "marginalizing actions and inertia of the university (organization)" that may be found within structures and practices that endorse a work climate hostile to members of underrepresented groups (Yosso et al 2009). More plainly, institutional microaggression is systematic racism through institutional rules and traditions. This systematic racism is carried out in hiring practices with built-in-biases, e.g. valuing candidates from Ivy league schools over HBCU/HSIs to 'diversify' faculty or valuing white women over WOC to 'diversity' faculty. In both cases, creating a policy that looks neutral but is subtly racist and/or sexist.

Gender Bias

As with racial microaggressions, gender bias can be subtle or implicit. Implicit gender bias is systemic and is often found in STEM departments of higher education and STEM workplaces. Gender bias includes gender gaps in salaries, workload differences, differences in promotion performance guideline adherence, and differential leadership opportunities. WOC face formidable barriers in these areas because gender bias exists on a personal level and also because biases are built into the organizational structures (Leggett-Robinson & Villa 2019). Spigel (2018) believes gender bias is partly due to androcentrism, the practice, conscious or unconscious, of placing a masculine point of view at the center of culture and history, thereby viewing women or any "other" as a deviation from normal. Based on previous research, the STEM workplace culture and climate is deeply embedded in an androcentric perspective and cultural norms in areas of salary, hiring, advancement, and leadership continue to marginalize WOC. (National Center of Education Statistics 2018, Spigel 2018, Williams 2006, Diangelo 2012).

For WOC in STEM, gender bias is particularly problematic. Not only must WOC surpass the "double oppression" of gender and race but they work in a white male dominated culture where the "glass ceiling," "chilling effect," and other manifestations of the "mobilization of the masculine bias" (Burton 1992) can be anticipated, regardless of the institution's or organization's established equity policies. Although several forms of gender bias occur within the STEM workplace, the most common forms are "prove-it again" and caregiving.

Prove-It-Again

Prove it again bias, is the idea that "in jobs historically held by men, men are presumed to be competent, while women often have to prove their competence over and over again. In addition, women's mistakes may be remembered forever, while men's are soon forgotten (Williams 2015).

Nearly 40 years of studies have documented that WOC in STEM fields are not seen as equally competent and are expected to perform at a higher level than their white male or non-male colleagues (Fiske and Taylor 2013; Foschi, 2000; Heilman 1983, 1984, 1995, 2012; Steele and Aronson 1995) and their successes are credited more to luck or circumstance than to skill. Based on the definition aforementioned, WOC in STEM are perceived as not fitting well within the bounds of the "the vision of a scientist," thus, they often have to provide more evidence of competence than do male (white or black) and white female scientists in order to be seen as equally competent (Eagly & Mladinic 1994; Foschi 1996; Foschi 2000). This descriptive stereotyping (Heilman, 2001) reflects the perceived lack of fit (Heilman 1983) between being a woman and being a scientist (Nosek, Banaji, & Greenwald 2002; Moss-Racusin et al 2012).

Prove-it-again bias leads WOC in STEM fields to work longer hours than their colleagues, resulting in a faltering work-life balance and an increase in health problems that may be stress-related (WIlliams, 2015). STEM fields provide fertile ground for "prove-it-again" gender bias for at least two reasons. First, studies of tokenism document that bias tends to occur more often when women make up less than 15% – 20% of a given field, which is common in many fields of science (although no longer in biology) (Kanter, 1977a, 1977b; Williams et.al 2014). Second, an influential study by Emilio Castilla and Stephen Benard found that bias is more common in fields like science, that see themselves as highly meritocratic (Castilla & Benard, 2010). One of the rare studies that compares white women in STEM with WOC in STEM found an important difference, namely that the white women reported higher levels of influence in their departments than did the WOC (Settles, Cortina, Malley, & Stewart, 2006). Another study found that WOC were likely (77%) to report that they felt they had to work harder than white women (58%) or white men (46%) to be perceived as legitimate (Hollenshead & Thomas, 2001; Williams, Phillips, & Hall, 2014). Prove-it-again bias is essentially a double standard for WOC in STEM and is manifested in activities that seek to justify lack of fit for women and promote in-group favoritism for men.

Caregiving

Caregiving bias (also known as *family responsibilities discrimination)* is the most commonly occuring but least discussed bias in the STEM workplace. Caregiving bias can be seen in a multitude of forms (Table 2); however, the form most visible is that of motherhood. Bias against mothers stems not only from assumptions about what mothers are like, but also from assumptions about how mothers should behave (Center for Worklife Law 2018). Additionally, this stereotype links motherhood with the lack of competence and perceived weakness in commitment to a career in STEM.

Ideal workers are dedicated to the job, meaning they are not supposed to take into consideration or be distracted by things that are not job-related (e.g., family or other outside responsibilities). Historically, ideal workers have been White men whose sole focus was on work. While all industries experience ideal-worker norms, the STEM culture is a level above. Research from Cech (2018) suggests the combination of expectations for long work hours above personal responsibilities like caregiving, make it particularly challenging for women with children, resulting in stymied career progress (Kulp, Wolf-Wendel, & Smith, 2019; Acker, 1990).

The intersectionality of gender and race through the lens of caregiving exposes a large gap in broadening and retaining participation from the perspective of family commitment, and has a strong impact on WOC in STEM. The conventional timing of earning a PhD, working in a post-doctoral position, entering the STEM workplace (academia, industry, or government) or climbing the STEM corporate ladder causes a substantial portion of career workload to occur during the critical decade between the ages of 30 and 40, also the time of life in which many women decide to start a family or have young children. The caregiving role can be particularly challenging

Table 2. Caregiving bias and consequences experienced by WOC in STEM

Bias	Consequence
Caregiving Challenges Among Early Career STEM Professionals	-Decrease in conference attendance -Decrease in professional development -Decrease in networking opportunities & professional advancement
Physical & Psychosocial Aspects of Childbearing Delay	-Infertility -Increase in the rates of anxiety and depression -Guilt and stigma -Relationship distress -Sexual dysfunction
Career Trajectories of New Parents in STEM	-Experience frustration, exhaustion, and disappointment -Difficulties managing career and family -Leading to a "pause" from career or leave the STEM field entirely
Cultural Expectations and Responsibilities of Caregiving	-Decrease relationship between family- and occupational-support -Increase work-family conflict and job satisfaction
Balancing Family and Dual Careers	-Occupational and family guilt due to competing obligations -Increased sense of exhaustion, guilt, and anxiety -Decreased time for self-care
Caregiving for Aging Parents	-Leading to a "pause" from career or leave the STEM field entirely -Experience interpersonal guilt -Economic and financial impact -Feelings of financial strain

for WOC who already struggle to create boundaries between work and home and are consumed with familial responsibilities. According to Mason (2010), the largest leak of WOC out of the STEM pipeline is when they start their families.

WOC in STEM fields compete on uneven ground because many institutional and organizational policies do not address the additional biases listed in Table 1. WOC will hear conflicting messages about her family's support for her professional career while she is expected to simultaneously continue in her role as caretaker and contributor to the family (Valenzuela, 2006). According to Schiebinger and Gilmartin (2010), female scientists do nearly twice as much work at home as their male counterparts, i.e., 54 percent of the cooking, cleaning, and laundry in their households whereby male scientists do just 28 percent, translating to an additional ten hours of work outside the demanding hours as a STEM professional. Additionally, due to their cultural obligations, WOC are often saddled with the responsibility of caring for aging parents (their own and spouse). In an attempt to avoid the "motherhood stigma" some WOC delay or avoid having children. This "bias avoidance" is well documented, especially in academia, as approximately 45 percent of tenured women are childless (Williams 2004). The "motherhood stigma" is in direct opposition to the daddy bonus in the workplace. Whereas, women with children are often penalized, the daddy bonus celebrates childbirth and men are paid more due to the preconceived "status" as the breadwinner for the family.

WOC experiencing microaggressions, isolation, and gender bias often have to contend with an additional layer of stress from their families in addition to the inherent challenges of succeeding in their desired STEM field. (Camacho & Lord, 2011; Carlone & Johnson, 2007; Ong, 2005). This added stress can adversely affect efforts by WOC of to build a cohesive identity in a STEM culture that does not consistently reflect or respect people who look like them (Smith et al., 2011; Wang et al., 2011). Counterspaces ("Safe STEM Ecosystems) offer WOC in STEM alternative engagement experiences, countering the distractions and adversities many face because of their racial/ethnic and gendered identities.

COUNTERSPACES AS "SAFE STEM ECOSYSTEMS"

In an effort to retain WOC in STEM, more attention must be paid to establishing safe social spaces, which offer support and enhance feelings of belonging in STEM. It is imperative to understand the formal and informal structures and relationships that support WOC, as well as enable them to contribute innovative knowledge (navigating the landscape) and perspectives in STEM for the next generation (Hess, Gault, & Yi, 2013). Several factors related to access and equity have created the need for counterspaces or "safe STEM ecosystems" for WOC. These factors (previously mentioned) are unwelcoming culture and climate, lack of STEM belonging, microaggressions, and gender bias.

Drawing on the work of Solórzano and his colleagues (Solórzano et al., 2000; Solórzano & Villalpando, 1998), *counterspaces* are defined as academic and social safe spaces that allow WOC to share their positive experiences; vent frustrations of isolation, microaggressions, and/or overt discrimination; and challenge deficit notions of WOC and, more importantly, establish and maintain a positive racial climate for themselves. In an effort to build the capacity of counterspaces for WOC in STEM, we have renamed counterspaces, appropriately, as "safe STEM Ecosystems". A STEM ecosystem is a large community of WOC linked together through experiences and common interests and goals. STEM ecosystems vary in size and seek to nurture and provide support for WOC at all stages of their STEM careers. Safe STEM Ecosystems are retreats from the "chilly" climate of STEM within their departments and/or organizations; these spaces lie in the margins, outside of mainstream educational or organizational spaces, and are occupied by

WOC in STEM fields. Safe STEM ecosystems are not comprised of "experts" but of WOC willing to share their victories and disappointments in order to assist others in navigating the STEM landscape.

Examples of counterspaces include but are not limited to peer-to-peer relationships, mentoring relationships, National STEM Diversity Conferences, and STEM/Non-STEM Campus groups. Unfortunately, these spaces do not readily exist at all STEM organizations whereby WOC can participate. As an alternative to the aforementioned counterspaces, *Critical Race Theory* research emphasizes the importance of empowering marginalized individuals to share their experiences in their own voices through the act of storytelling. (DeCuir-Gunby & Walker-DeVose, 2013; Delgado & Stefancic, 2012; Leyva, 2016; Solórzano & Villalpando, 1998; Yosso, 2006).

Storytelling allows WOC to describe their experiences of discrimination and oppression in safe STEM Ecosystems.

STORYTELLING IN STEM

The society we live in today could not exist without the influence from past generations. Human beings, as creatures of habit and tradition, hand down lessons and knowledge from one generation to the next. Storytelling has been a traditional and even ancient means of passing on wisdom and culture (Brady & Haley, 2013). It allows us to leave a legacy of knowledge for our descendants to utilize while connecting us to our humanity, linking us to our past, and providing a glimpse into our future (Freed 2011).

Our ability to learn from our ancestors' past decisions and use that knowledge to give ourselves an insight towards our own problems provides an advantage rarely found in other species, that of a communal knowledge pool. The tradition of storytelling facilitates this transfer of knowledge. Stories teach us to love, to forgive others, to be just and to strive for better than we have. As a result, storytelling remains to this day the single most important tradition humans participate in. In the Black and Brown community, storytelling is powerful. It is about validation, a way of saying *I am here and I matter and bearing witness to what one has been through.*

The same is true in the stories about the STEM culture from WOC in STEM. These stories function to also pass down lessons, information, and inspiration about the STEM landscape. Through storytelling, WOC in STEM contribute to the aforementioned ecosystems through shared experiences and comfort knowing they are not "alone" on the STEM journey. It gives a voice that they matter, regardless of how the narrative was created and disseminated and decreases feelings of isolation and self-presentation. Storytelling aligned with tools, plans, validation, and encouragement gives WOC the power to replace the unimaginable weight of being made to feel invisible with the positivity of their voice.

Unfortunately, as WOC advance in STEM, their stories become more difficult to share and there is a tendency to speak more impersonally about the STEM landscape or to not fully disclose their disappointments and hurts, for fear of appearing weak, angry, bitter, pessimistic or ungrateful. The pain associated with memories of microaggressions, isolation, and gender bias of the STEM landscape is difficult to manage. However, Glen Singleton (2015) reminds us "that speaking your truth about race (in this case the intersectionality of race, gender, and STEM) requires a willingness to take risks". It means being honest about our thoughts, feelings, and opinions as it relates to the landscape of STEM fields.

Through the counterspaces of storytelling - STEM Autobiographies, these difficult truths can be unpacked, exposed, and examined in spaces that provide refuge, credence, and inspiration. STEM Autobiographies are tools used in storytelling by WOC that provide a process for developing and deepening personal understanding of the transformations of navigating the landscape.

A STEM AUTOBIOGRAPHY

I am an African-American woman, the oldest of four children from a blended family. Two children grew up in one house and two grew up in another house. Both homes, although in two different states - Georgia and Florida - were in predominantly middle class African American neighborhoods. Both homes were also located in large cities in the south, Atlanta and Miami. In both homes, education was considered the way to upward mobility and it was emphasized, *strongly*.

As I reach back into my memory to recall the earliest experience of intersectionality in STEM, my recollection takes me all the way back to high school. Initially, I thought of my most recent professional experience as being the first encounter, because my graduate school experience had been amazing and I had simply characterized other "incidents" as normal educational and professional growing pains; however, I now realize I have several memories where my intersectionality of race and gender in STEM affected how I was viewed by the culture and how I responded.

In graduate school, I didn't encounter issues of stereotyping, gender bias, microaggressions, or the like. I had a great advisor who not only believed in STEM diversity, equity, and inclusion (DEI) in the early 2000's, but practiced DEI throughout his entire higher education administrative career. Our research lab consisted of Asian, Hispanic, and African-American students, with the majority being women; the research associate was also Hispanic. The year I graduated with my Ph.D., there were 9 others graduating with a Ph.D. in chemistry with the same racial, ethnic, and gender profile as myself. Not once did I view this experience as being "out of the norm". The ten of us were a close knit group. We discussed the challenges of obtaining a Ph.D. in chemistry; however, I don't recall any conversations of us having to confront the nastiness of microaggressions, discrimination, or gender bias. In retrospect, if we had, I am not sure if we actually had the language to discuss systematic issues.

High School Education

High school was actually my earliest memory of being challenged while navigating the STEM landscape. I was "bussed" to a school outside my neighborhood, where the majority of the students were white. The "M to M" program was designed to give minority students the same academic experience of majority students. Although my elementary school (grades 1st-8th) had some white kids, there were so few, we could count and name them on *two* hands. My day started at 5:30 am, and I would catch two buses in order to get to school on time. The county provided the transportation and it was fun riding with people from neighborhoods similar in demographics. Once we arrived at school, I was lucky to have 3 or four students that "looked like me" in the same class. Nonetheless, I learned to adapt and make friends. At 13 years old, I rationalized that by attending the all-white high school, I could have more friends, e.g. friends at school, friends on the bus, and friends in my neighborhood. Academic excellence came easy to me. I loved school and enjoyed studying. I recall my mother ensuring I took the most rigorous college preparatory classes. Back then, she realized the importance of taking STEM courses and preparing for a STEM career.

During my senior year, I registered for a Physics class. Up until that time, I had enjoyed all of my STEM classes and experienced success in each of them. Chemistry was my favorite - I loved the discovery of chemical reactions and often received detention for performing chemistry experiments "outside" the curriculum. My math classes were challenging but I still deemed them as fun. My physics class was challenging because of the material, but also because I struggled to engage with my teacher. She was a retired physics professor from the University of Alabama and constantly spoke in a belittling tone to me and the other 2 black students in the class. As my recollection takes me back some 33 years ago, I can still hear her voice "maybe you should take another class that doesn't require so much math. Ya'll just typically aren't that good at it". Coupled with her negative comments were her refusal to assist us or answer our questions. Her comments and disengagement affected both my emotional and physical state (triggered my eczema). Although I (along with the other two black girls) left her class, each of us struggled in math in our first and second year of college. At 16, I had no idea that this teacher was placing the barriers of stereotyping, racial, and gender bias on my STEM path.

Undergraduate Education

I attended a large R1 institution for my undergraduate education. My major was pharmacy which meant the core courses were chemistry, math, and physics. Once again, however, I struggled in math. Each time I received a failing score, I kept thinking, perhaps the physics teacher was correct, especially since I could count on one hand the number of students of color (male and female) taking the class out of a class of ~200. To add insult to injury, I also had difficulties passing the math portion of the chemistry entrance exam. After the 3rd try, I passed. But with each fail, I was faced with the stark reality that perhaps I just couldn't do math. However, it was not until I enrolled in my general biology lab class that I encountered the "chilly climate" of STEM.

The first day of class, we separated into lab partners. I was the only person of color (African-American) in the lab class (30 people). I remember clearly the class dividing up into pairs and threes and being the only person left without a partner. I thought it strange, but never commented. I worked the entire semester by myself. The second semester was a repeat of the first, but by that time, I was used to working and figuring things out by myself.

I ended up leaving the institution because of my poor academic start in math and chemistry. By the time I got to physics, I realized the institution was not the place for me. I excelled socially but in my STEM coursework, I literally failed. Upon returning home, I attended a community college and then transferred to an urban institution. My experiences at the urban institution were nurturing, but I was also gaining an understanding of what it was like to be a WOC in STEM. My calculus professor (adjunct from Spelman) took the time to remind me that I was competent and capable of excelling in mathematics. But just as I was reminded by him, I was subjected to the microaggressions from the chemistry professors with comments of "Wow, I can't believe you scored so well on the professional exam. Most people like you don't do well on standardized tests" or "I am glad to see you are good in math. Most girls like you aren't, so it is good for you, being a black girl. You will do well." When I did complain to the chair, the response was that the professor was older and probably didn't realize what he was saying. Regardless if he did or not, I was left feeling as though I was doing something that "other black students" couldn't do. It was actually these types of comments that lead me to a career in the professoriate. I wanted to "help" other black students become better at math and chemistry and the best way was through being a professor.

The Professoriate

Life in the academy was not "difficult" per se. It was challenging, but I knew the "game": get great teaching evaluations, do service, and accumulate scholarly achievements. I had become an overachiever (once I transferred to the urban institution), so I knew exactly what to do. I navigated the tenure/promotion process flawlessly and actually was able to separate out my promotion from tenure. I was on the fast track to an administrative position in higher education and I was ready.

It was during my climb to a full professor from associate professor that I began experiencing difficulties in navigating the STEM landscape. As a STEM administrator, I faced microaggressions from both my colleagues and eventually my supervisor. During my tenure as an administrator, I also took notice of the constant practice of tokenism on DEI committees and the lack of inclusion of myself on committees that were substantively significant i.e. hiring committees, consolidation committees, etc. Most of the time however, I would choose to ignore the comments and rationalize them as something other than what they actually were; a common survival skill for myself and perhaps many other WOC in STEM.

Although I was the recipient of these microaggressions, I coped by overachieving. I figured they would see that I deserved to be placed on the "significant" committees. I soon realized my achievements did not matter when I was passed over for a promotion. The promotion was given to a male with the same administrative experience but significantly less grant funding success and scholarly activity. I was devastated and had difficulty rebounding. The situation continued to dive as my previous "peers" were now my supervisors. I was tasked with expectations that my male counterparts did not have and was constantly questioned about my departmental decisions. These behaviors began to take effect on my health and my hair began to break. My hair was in such bad condition I had to cut it off and began wearing it in a natural style (previously, I wore my hair straight, more European). The remarks from my supervisor regarding my "new" hair were not flattering. In fact, the remarks made me question if I even belonged in the division. For the first time, I actually realized, I was the victim of microaggressions and gender bias. Upon meeting with an executive administrator regarding these acts, there was inaction, retaliation, and a letter of constructive resignation. I left the one place I felt I could make a difference for other people that looked like me.

Celebrating the Transformation

After some time, I was able to re-group via a safe STEM space (two WOC). They helped me to see that although what I had encountered, there were several other WOC that had suffered just as badly as I had. These women had also lost their "dream", but in the midst of losing, their lives (personal and professional) were transformed into something far better. I was tasked to cry and cry until I was out of tears. I was encouraged to seek help via counseling. I was forced to assess my strengths and weaknesses as well as define myself ***without*** a job title. Being a WOC in STEM, resulted in my clinging to every title, award, and role that made me feel a sense of belonging and value in a broken system. These accolades became my identity, ultimately resulting in self-dehumanization (Bastian et.al, 2013; Scheier & Carver, 1983; Bastian & Haslan, 2011). I had adopted a self-perception of a successful woman of color in STEM but also a meta-perception whereby I saw myself through the eyes of the STEM culture and had adopted the STEM culture bias of "prove-it again" for myself and everyone I mentored. Once I was able to do the "assigned homework" and come to grips with the realities of self-humanization, I was able to look at my journey from a very different perspective.

My journey is similar to that of a butterfly - moving from one dimension to another, from knowing something on the surface, then undergoing a deep experience that leads to an entirely new self-capacity and perception. Through this journey I found strength in myself and learned to lean on those WOC in STEM that paved the way (through their experiences). My transformational journey has created safe spaces for WOC in STEM to share their stories, to have their voices heard, and to feel empowered.

I celebrate every barrier experienced while navigating the STEM landscape. The process of overcoming each barrier I've faced has made me more resilient, more vocal, and more self-aware of the social injustices within STEM. I am no longer afraid to take the risk in telling my story or stories of other WOC in STEM. I am no longer embarrassed to shed tears about my journey.

Each day I wake up with a mission similar to the one developed during my undergraduate years, twenty-six years ago. Although more pointed and far reaching, it focuses on "helping" WOC navigate the STEM landscape. I carry out this mission by serving as a consultant for WOC in STEM and creating safe STEM ecosystems (programs and storytelling) in which these historically marginalized voices can be heard and feel a sense of belonging and validation. WIth each program and story, WOC are empowered to *embrace* their gender and race within the STEM culture.

REFERENCES

Acker, J. (1990). Hierarchies, Jobs, Bodies: A Theory of Gendered Organizations. *Gender & Society*, *4*(2), 139–158. doi:10.1177/089124390004002002

Alfred, M. V., Ray, S. M., & Johnson, M. A. (2019). Advancing Women of Color in STEM: An Imperative for U.S. Global Competitiveness. *Advances in Developing Human Resources*, *21*(1), 114–132. doi:10.1177/1523422318814551

Allen-Ramdial, S. A., & Campbell, A. G. (2014). Reimagining the Pipeline: Advancing STEM Diversity, Persistence, and Success. *Bioscience*, *64*(7), 612–618. doi:10.1093/biosci/biu076 PMID:25561747

Bastian, B., Jetten, J., Chen, H., Radke, H. R. M., Harding, J. F., & Fasoli, F. (2013). Losing our humanity: The self-dehumanizing consequences of social ostracism. *Personality and Social Psychology Bulletin*, *39*(2), 156–169. doi:10.1177/0146167212471205 PMID:23386654

Bastian, B., & Haslam, N. (2011). Experiencing dehumanization: Cognitive and emotional effects of everyday dehumanization. *Basic and Applied Social Psychology*, *33*(4), 295–303. doi:10.1080/01973533.2011.614132

Baumeister, R. F., & Leary, M. R. (1995). The need to belong: Desire for interpersonal attachments as a fundamental human motivation. *Psychological Bulletin*, *117*(3), 497–529. doi:10.1037/0033-2909.117.3.497 PMID:7777651

Bourdieu P. & Collier, P. (1988) *Homo Academicus*. Cambridge: Polity Press in association with Basil Blackwell.

Brady, W. H., & Haley, S. (2013). Storytelling defines your organizational culture. *Physician Executive*, *39*(1), 40–43. PMID:23437755

Brown, S. V. (2000). The preparation of minorities for academic careers in science and engineering: How well are we doing? In G. Campbell, R. Denes, & C. Morrison (Eds.), *Access denied: Race, ethnicity, and the scientific enterprise*. New York: Oxford University Press.

Brown, S. W. (2008). The gender differences: Hispanic females and males majoring in science or engineering. *Journal of Women and Minorities in Science and Engineering*, *14*(2), 205–223. doi:10.1615/JWomenMinorScienEng.v14.i2.50

Burton, C. (1992). Merit and gender: Organizations and the mobilization of masculine bias. In A. Mills & P. Tancred (Eds.), *Gendering organizational analysis*. Newbury Park, CA: SAGE Publishing.

Camacho, M. M., & Lord, S. M. (2011). *"Microaggressions" in engineering education: Climate for Asian, Latina and White women.* Paper presented at Frontiers in Education Conference (FIE), Rapid City, SD. 10.1109/FIE.2011.6142970

Cárdenas, V., & Treuhaft, S. (2013), All-In Nation: An America that Works for All. *Center for American Progress*. https://www.americanprogress.org/issues/race/reports/2013/07/24/70540/all-in-nation-an-america-that-works-for-all/

Carlone, H. B., & Johnson, A. (2007). Understanding the science experiences of successful women of color: Science identity as an analytic lens. *Journal of Research in Science Teaching*, *44*(8), 1011–1245. doi:10.1002/tea.20237

Castilla, E., & Bernard, S. (2010). The paradox of meritocracy in organizations. *Administrative Science Quarterly*, *55*(4), 543–676. doi:10.2189/asqu.2010.55.4.543

Center for Worklife Law. (2018). *Maternal Wall -Gender Bias Learning Project.* Retrieved from: https://genderbiasbingo.com/maternal-wall/

Charleston, L. J., Adserias, R. P., Lang, N. M., & Jackson, J. F. L. (2014). Intersectionality and STEM: The role of race and gender in the academic pursuits of African American women in STEM. *Journal of Progressive Policy and Practice*, *2*(3), 273–293. Retrieved from https://caarpweb.org/wp-content/uploads/2014/12/Charleston-Adserias-Lang-Jackson-2014.pdf

Cress, C. M., & Sax, L. J. (1998). Campus climate issues to consider for the next decade. *New Directions for Institutional Research*, *98*(98), 65–80. doi:10.1002/ir.9805

Eagly, A., & Johnson, B. (1990). Gender and leadership style: A meta-analysis. *Psychological Bulletin*, *108*(2), 233–256. doi:10.1037/0033-2909.108.2.233

DeAngelis, T. (2009). Unmasking "racial microaggressions.". *Monitor on Psychology*, *40*, 42–46.

DeCuir-Gunby, J. T., Johnson, O. T., Edwards, C. W., McCoy, W. N., & White, A. M. (2019). African American professionals in higher education: Experiencing and coping with racial microaggressions. *Race, Ethnicity and Education*, 1–17. doi:10.1080/13613324.2019.1579706

DeCuir-Gunby, J. T., & Walker-DeVose, D. C. (2013). Expanding the counter-story: The potential for critical race mixed methods studies in education. In M. Lynn & A. D. Dixson (Eds.), *Handbook of critical race theory in education*. New York, NY: Routledge.

Delgado, R., & Stefancic, J. (2012). *Critical race theory: An introduction* (2nd ed.). New York: New York University Press.

Diangelo, R. (2012). *What Does It Mean to Be White? Developing White Racial Literacy*. New York, NY: Peter Lang.

Fiske, S., & Taylor, S. (2013). *Social Cognition: From Brains to Culture*. Thousand Oaks, CA: SAGE Publications. doi:10.4135/9781446286395

Fleming, J. (1984). *Blacks in College: A Comparative Study of Students' Success in Black and White Institutions*. San Francisco: Jossey-Bass.

Foschi, M. (1996). Double standards in the evaluation of men and women. *Social Psychology Quarterly*, *59*(3), 237–254. doi:10.2307/2787021

Foschi, M. (2000). Double Standards for Competence: Theory and Research. *Annual Review of Sociology*, *26*(1), 21–42. doi:10.1146/annurev.soc.26.1.21

Freed, R. (2011, November 17). The Importance of Telling Our Stories [Blog post]. Retrieved from https://www.huffpost.com/entry/legacy-telling-our-story_b_776195

Good, C., Rattan, A., & Dweck, C. S. (2012). Why do women opt out? Sense of belonging and women's representation in mathematics. *Journal of Personality and Social Psychology*, *102*(4), 700–717. doi:10.1037/a0026659 PMID:22288527

Goodenow, C. (1993). Classroom belonging among early adolescent students: Relationships to motivation and achievement. *The Journal of Early Adolescence*, *13*(1), 21–43. doi:10.1177/0272431693013001002

Heilman, M. (1983). Sex bias in work settings: The lack of fit model. *Research in Organizational Behavior*, *5*, 269–986.

Heilman, M. (1984). Information as a deterrent against sex discrimination: The effects of applicant sex and information type on preliminary employment decisions. *Organizational Behavior and Human Performance*, *33*(2), 174–186. doi:10.1016/0030-5073(84)90019-9

Heilman, M. (2001). Description and prescription: How gender stereotypes prevent women's ascent up the organizational ladder. *The Journal of Social Issues*, *57*(4), 657–674. doi:10.1111/0022-4537.00234

Heilman, M. (2012). Gender Stereotypes and Workplace Bias. *Research in Organizational Behavior*, *32*, 113–135. doi:10.1016/j.riob.2012.11.003

Hess, C., Gault, B., & Yi, Y. (2013). *Accelerating change for women of color faculty in STEM: Policy, action, and collaboration*. Retrieved from https://iwpr.org/publications/accelerating-change-for-women-faculty-of-color-in-stem-policy-action-and-collaboration/

Hill, B., Secker, J., & Davidson, F. (2014). Achievement Relative to Opportunity: Career Hijacks in the Academy. *Advances in Gender Research*, *19*, 85–107. doi:10.1108/S1529-212620140000019004

Hollenshead, C., & Thomas, G. (2001). Resisting from the Margins: The Coping Strategies of Black Women and Other Women of Color Faculty Members at a Research University. *The Journal of Negro Education*, *70*(3), 166–175. doi:10.2307/3211208

Ireland, D. T., Freeman, K. E., Winston-Proctor, C. E., DeLaine, K. D., McDonald Lowe, S., & Woodson, K. M. (2018). (Un)Hidden Figures: A Synthesis of Research Examining the Intersectional Experiences of Black Women and Girls in STEM Education. *Review of Research in Education*, *42*(1), 226–254. doi:10.3102/0091732X18759072

Johnson, A., Brown, J., Carlone, H., & Cuevas, A. K. (2011). Authoring identity amidst the treacherous terrain of science: A multiracial feminist examination of the journeys of three women of color in science. *Journal of Research in Science Teaching*, *48*(4), 339–366. doi:10.1002/tea.20411

Johnson, D. R. (2011). Examining sense of belonging and campus racial diversity experiences among women of color in STEM Living-Learning Programs. *Journal of Women and Minorities in Science and Engineering*, *17*(3), 209–223. doi:10.1615/JWomenMinorScienEng.2011002843

Johnson, D. R. (2012). Campus racial climate perceptions and overall sense of belonging among racially diverse women in STEM majors. *Journal of College Student Development*, *53*(2), 336–346. doi:10.1353/csd.2012.0028

Kachchaf, R., Ko, L., Hodari, A., & Ong, M. (2015). Career-life balance for women of color: Experiences in science and engineering academia. *Journal of Diversity in Higher Education*, *8*(3), 175–191. doi:10.1037/a0039068

Kang, C. (2019, June 5). Turning up the flow won't fix the STEM pipeline problem [Blog post & figure]. Retrieved from https://medium.com/race-law-a-critical-analysis/turning-up-the-flow-wont-fix-the-stem-pipeline-problem-4166135dd807

Kanter, R. M. (1977a). Some effects of proportions on group life: Skewed sex ratios and responses to token women. *American Journal of Sociology*, *85*(5), 965–990. doi:10.1086/226425

Kanter, R. M. (1977b). *Men and women of the corporation*. New York: Basic Books.

Kerr, N. L., & Levine, J. M. (2008). The detection of social exclusion: Evolution and beyond. *Group Dynamics*, *12*(1), 39–52. doi:10.1037/1089-2699.12.1.39

Ko, L. T., Kachchaf, R. R., Hodari, A. K., & Ong, M. (2014). Agency of women of color in physics and astronomy: Strategies for persistence and success. *Journal of Women and Minorities in Science and Engineering*, *20*(2), 171–195. doi:10.1615/JWomenMinorScienEng.2014008198

Kulp, A., Wolf-Wendel, L., & Smith, D. (2019). The Possibility of Promotion: How Race and Gender Predict Promotion Clarity for Associate Professors. *Teachers College Record*, *121*(5).

Lawrence, K., & Keleher, T. (2004). Structural Racism. *Race and Public Policy Conference Proceedings*.

Leggett-Robinson, P., Villa, B., & Davis, N. (2018). Cultivating STEM Identity & Belonging through Civic Engagement: Increasing Student Success (Self-Efficacy and Persistence) for the Two-year college STEM student. *Science Education and Civic Engagement: An International Journal, 10*(1), 23-34. http://new.seceij.net/wp-content/uploads/2018/03/Leggett.pdf

Leggett-Robinson, P., & Villa, B. (2019) Navigating the Landscape of the STEM Professoriate: Reflections and Insights From Women of Color. In *Women's Influence on Inclusion, Equity, and Diversity in STEM Fields.* IGI Global Publishing. https://www.igi-global.com/book/women-influence-inclusion-equity-diversity/218525

Leyva, L. A. (2016). An intersectional analysis of Latin@ college women's counter-stories in mathematics. *Journal of Urban Mathematics Education*, *9*(2), 81–121. Retrieved from http://ed-osprey.gsu. edu/ojs/index.php/JUME/article/view/295

Lewis, K., Stout, J. G., Pollock, S. J., Finkelstein, N. D., & Ito, T. A. (2016). Fitting in or opting out: A review of key social-psychological factors influencing a sense of belonging for women in physics. *Physical Review Physics Education Research*, *12*(2), 020110. doi:10.1103/PhysRevPhysEducRes.12.020110

Lezotte, S., & Ferguson, S. (2017). Envisioning Women in STEM. In *Feminist Activist Symposium*. Glassboro, NJ: Rowan University.

MacPherson, W. (1999). *The Stephen Lawrence Inquiry*. London: The Stationery Office.

Mason, M.A, Goulden, M., & Frasch, K. (2010). Keeping women in the science pipeline. *Proceedings from Workplace Flexibility Georgetown Law: Focus on Workplace Flexibility*.

Moss-Racusin, C., Dovidio, J., Brescoll, V., Graham, M., & Handelsman, J. (2012). Faculty's subtle gender biases favor male students. *Proceedings of the National Academy of Sciences of the United States of America, 109*(41), 16474–16479. doi:10.1073/pnas.1211286109 PMID:22988126

National Academy of Sciences, National Academy of Engineering, and Institute of Medicine. (2011). *Expanding Underrepresented Minority Participation: America's Science and Technology Talent at the Crossroads*. Washington, DC: The National Academies Press. doi:10.17226/12984

National Academies of Sciences, Engineering, and Medicine. (2016). *Barriers and Opportunities for 2-Year and 4-Year STEM Degrees: Systemic Change to Support Diverse Student Pathways*. Committee on Barriers and Opportunities in Completing 2-Year and 4-Year STEM Degrees. S. Malcom and M. Feder, Editors. Board on Science Education, Division of Behavioral and Social Sciences and Education. Board on Higher Education and the Workforce, Policy and Global Affairs. Washington, DC: The National Academies Press. doi:10.17226/21739

National Center of Education Statistics (NCES). (2018). *The Condition of Education 2018*. Washington, DC: Author.

National Center for Education Statistics (NCES). (2013). *Projections for Education Statistics to 2021*. https://nces.ed.gov/programs/projections/projections2021/sec2b.asp

National Center for Education Statistics (NCES). (2017a). Integrated Postsecondary Education Data System (IPEDS), Completions component. In *Digest of Education Statistics: 2016 Tables and Figures.* Author.

National Center for Education Statistics (NCES). (2017b). Number and Percentage Distribution of Science, Technology, Engineering, and Mathematics (STEM) Degrees/Certificates Conferred by Postsecondary Institutions, by Race/Ethnicity, Level of Degree/Certificate, and Sex of Student: 2008-09 through 2015-16. In *Digest of Education Statistics: 2016 Tables and Figures*. Author.

National Research Council (NRC). (2012). *Community colleges in the evolving stem education landscape: Summary of a summit*. Washington, DC: National Academies Press.

National Science Foundation (NSF), National Center for Science and Engineering Statistics. (2017). *Women, Minorities, and Persons with Disabilities in Science and Engineering: 2017.* Special Report NSF 17-310. Available at www.nsf.gov/statistics/wmpd/

Nosek, B. A., Banaji, M. R., & Greenwald, A. G. (2002). Math = Male, Me = Female, Therefore Math ≠Me. *Journal of Personality and Social Psychology, 83*(1), 44–59. doi:10.1037/0022-3514.83.1.44 PMID:12088131

O'Conner, L. T., & Cech, E. (2018). Not Just a Mothers' Problem: The Consequences of Perceived Workplace Flexibility Bias for All Workers. *Sociological Perspectives, 61*(5), 808–829. doi:10.1177/0731121418768235

Ong, M. (2001). Playing with invisibility: How minority women gain power from the margins of science culture. *Women in Higher Education, 10*(11), 4–44.

Ong, M. (2005). Body projects of young women of color in physics: Intersections of gender, race, and science. *Social Problems, 52*(4), 593–617. doi:10.1525p.2005.52.4.593

Ong, M. (2011). Broadening participation: The status of women of color in computer science. *Communications of the ACM, 54*(7), 32–34. doi:10.1145/1965724.1965737

Ong, M., Smith, J. M., & Ko, L. T. (2018). Counterspaces for women of color in STEM higher education: Marginal and central spaces for persistence and success. *Journal of Research in Science Teaching, 5*(2), 206–245. doi:10.1002/tea.21417

Parker, A. D. (2013). *Family matters: Familial support and science identity formation for African American female STEM majors* (Doctoral dissertation). Retrieved from ProQuest Dissertations and Theses database. (Publication No. AAT 3594079)

Paludi, M. (2008). The Psychology of Women at Work: Challenges and Solutions for Our Female Workforce. *Praeger Perspectives, Women's psychology*. ABC-CLIO.

Rainey, K., Dancy, M., Mickelson, R., Stearns, E., & Moller, S. (2018). Race and gender differences in how sense of belonging influences decisions to major in STEM. *International Journal of STEM Education, 5*(10), 10. doi:10.118640594-018-0115-6 PMID:30631700

Scheier, M. F., & Carver, C. S. (1983). Self-directed attention and the comparison of self with standards. *Journal of Experimental Social Psychology, 19*(3), 205–222. doi:10.1016/0022-1031(83)90038-0

Schiebinger, L., & Gilmartin, S. (2010). Housework Is an Academic Issue. *Academe, 96*(1), 39–44.

Sethna, B. N. (2011). Minorities in higher education: A pipeline problem? *Research in Higher Education, 13*, 1–18.

Settles, I., Cortina, L., Malley, J., & Stewart, A. (2006). The climate for women in academic science: The good, the bad, and the changeable. *Psychology of Women Quarterly*, *30*(1), 47–58. doi:10.1111/j.1471-6402.2006.00261.x

Shehab, R., Murphy, T., Davidson, J., Foor, C., Reed Roads, T., Trytten, T., & Walden, S. (2007). *Experiences as a non-majority engineering student.* Paper presented at the American Society for Engineering Education Annual Conference & Exposition, Atlanta, GA.

Singleton, G. E. (2015) Courageous Conversations about Race: a Field Guide for Achieving Equity in Schools. Thousand Oaks, CA: Corwin, A SAGE Company.

Smith, J. L., Lewis, K. L., Hawthorne, L., & Hodges, S. D. (2013). When trying hard isn't natural: Women's belonging with and motivation for male-dominated STEM fields as a function of effort expenditure concerns. *Personality and Social Psychology Bulletin*, *39*(2), 131–143. doi:10.1177/0146167212468332 PMID:23187722

Smith, W. A., Hung, M., & Franklin, J. D. (2011). Racial battle fatigue and the miseducation of Black men: Racial microaggressions, societal problems, and environmental stress. *The Journal of Negro Education*, *80*(1), 63–82. Retrieved from http://www.journalnegroed.org/recentissues.html

Snyder, K. (2014, October 2) Why Women Leave Tech: It's the Culture, Not Because 'Math Is Hard' [Blog Post]. fortune.com/2014/10/02/women-leave-tech-culture/.

Solórzano, D., & Villalpando, O. (1998). Critical race theory, marginality, and the experience of minority students in higher education. In C. Torres & T. Mitchell (Eds.), *Emerging issues in the sociology of education: Comparative perspectives.* Albany: State University of New York Press.

Solórzano, D., Ceja, M., & Yosso, T. (2000). Critical race theory, racial microaggressions, and campus racial climate: The experiences of African American college students. *The Journal of Negro Education*, *69*(1), 60–73. Retrieved from http://www.jstor.org/stable/2696265

Spigel, R. (2018, June 25). Challenging Androcentrism and Implicit Bias in the Academy [Blog Post]. Academic Impressions https://www.academicimpressions.com/blog/androcentrism-implicit-bias-higher-education/

Steele, C., & Aronson, J. (1995). Stereotype threat and the intellectual test performance of African Americans. *Journal of Personality and Social Psychology*, *69*(5), 797–811. doi:10.1037/0022-3514.69.5.797 PMID:7473032

Steele, C., Spencer, S., & Aronson, J. (2002). Contending with group image: The psychology of stereotype and social identity threat. *Advances in Experimental Social Psychology*, *34*, 379–440. doi:10.1016/S0065-2601(02)80009-0

Sue, D. W., Bucceri, J., Lin, A. I., Nadal, K. L., & Torino, G. C. (2007a). Racism Microaggressions and the Asian American Experience. *Cultural Diversity & Ethnic Minority Psychology*, *13*(1), 72–81. doi:10.1037/1099-9809.13.1.72 PMID:17227179

Valenzuela, Y. (2006). *Mi fuerza/my strength: The academic and personal experiences of Chicana/Latina transfer students in math and science* (Doctoral dissertation). Retrieved from ProQuest Dissertation and Theses database. (Publication No. 304916976)

Varma, R. (2002). Women in information technology: A case study of undergraduate students in a minority-serving institution. *Bulletin of Science, Technology & Society*, *22*(4), 274–282. doi:10.1177/0270467602022004003

Varma, R., Prasad, A., & Kapur, D. (2006). Confronting the "socialization" barrier: Crossethnic differences in undergraduate women's preference for IT education. In J. M. Cohoon & W. Aspray (Eds.), *Women and information technology: Research on underrepresentation*. Cambridge, MA: MIT Press. doi:10.7551/mitpress/9780262033459.003.0010

Wang, J., Leu, J., & Shoda, Y. (2011). When the seemingly innocuous "stings": Racial microaggressions and their emotional consequences. *Personality and Social Psychology Bulletin*, *37*(12), 1666–1678. doi:10.1177/0146167211416130 PMID:21885859

Ward, K., & Wolf-Wendel, L. (2012). *Academic motherhood: How faculty manage work and family*. New Brunswick, NJ: Rutgers University Press.

Ward, K., & Wolf-Wendel, L. (2016). Academic motherhood: Mid-career perspectives and the ideal worker norm. In L. Wolf-Wendel, K. Ward, & A. Kulp (Eds.), *How ideal worker norms shape work-life balance for different constituent groups in higher education (New Directions in Higher Education, No. 176)*. San Francisco, CA: JosseyBass. doi:10.1002/he.20206

Wesselmann, E. D., Cardoso, F. D., Slater, S., & Williams, K. D. (2012). To be looked at as though air: Civil attention matters. *Psychological Science*, *23*(2), 166–168. doi:10.1177/0956797611427921 PMID:22246319

Whittaker, J. A., & Montgomery, B. A. (2014). Cultivating Institutional Transformation and Sustainable STEM Diversity in Higher Education through Integrative Faculty Development. *Innovative Higher Education*, *39*(4), 263–275. doi:10.100710755-013-9277-9

Williams, A., Alon, T., & Bornstein, S. (2006, Fall). Beyond the 'Chilly Climate': Eliminating Bias Against Women and Fathers in Academe. *Thought & Action*, 79–96.

Williams, J. (2004). Hitting the Maternal Wall-Before They Reach a "Glass Ceiling" in Their Careers, Women Faculty May Hit a "Maternal Wall". *Academe Bulletin of the American Association of University Professors*, *90*, 16.

Williams, J. (2015). The 5 Biases Pushing Women Out of STEM. *Harvard Business Review*. https://hbr.org/2015/03/the-5-biases-pushing-women-out-of-stem

Williams, J. C., Phillips, K. W., & Hall, E. V. (2014). Double Jeopardy? Gender Bias Against Women in Science. *The Center for WorkLife Law*. https://worklifelaw.org/publications/Double-Jeopardy-Report_v6_full_web-sm.pdf

Yosso, T. J. (2006). *Critical race counterstories along the Chicana/Chicano educational pipeline*. New York: Routledge.

Yosso, T. J., Smith, W. A., Ceja, M., & Solórzano, D. G. (2009). Critical race theory, racial microaggressions, and campus racial climate for Latina/o undergraduates. *Harvard Educational Review*, *79*(4), 659–691. doi:10.17763/haer.79.4.m6867014157m7071

Chapter 4
Successful STEM Women of Color Must Network Differently

Sherine O. Obare
University of North Carolina at Greensboro, USA & North Carolina A&T State University, USA

ABSTRACT

Overcoming barriers for women of color in STEM can be eased not only by what you know, but rather by who you know. Women are always encouraged to network, but often are not given the full instructions on what the outcome of the network could or should be. Seldom are they mentored to network in ways that they are viewed as contributors who bring impact and value to the discipline. The purpose of this chapter is to provide some guidelines from lessons learned about networking not only through the author's experience as a faculty member moving through the ranks, but as a university administrator, and a leader in a national organization. The chapter will have examples of various ways that women of color can contribute through networking to create opportunities to make significant impact to their discipline. The chapter will also reference some key books and articles that share some of the most pertinent work being done.

INTRODUCTION

Importance of Networking

Women of color who select to pursue careers in academia are faced with a myriad of barriers with a major one being isolation in their departments and their disciplines. Such isolation makes it difficult to navigate the tenure and promotion arena,

DOI: 10.4018/978-1-7998-4858-5.ch004

opportunities for professional growth, and leadership roles. Universities always state that you have to excel in research, teaching, and service. However, the obstacles that one must overcome to excel in each of these areas, while also trying to balance your personal life is a difficult task.

It took me many years to understand how to network effectively and how to ensure that the relationships I was building were mutually beneficial to my professional career. I must admit, going through various aspects of my doctoral and postdoctoral training prepared me well to be a great researcher, but not to be a "networker". I learned a lot through the years by trial and error. At times I felt that I knew what I was doing, and at times I had to figure it out. I hope to share some of my experiences in this chapter with the goal that the reader will find it easier to navigate the professoriate by being a strong and more equipped networker. It took me time to understand good practices for networking, and much of what I learned came from the literature by Bradford (2014); Coburn (2014); de Janasz et al. (2008); Fisher et al. (2000); Lippe-McGraw (2019); Marcus (2019) and Spurk et al. (2015), as well as personal interactions.

Figure 1 shows a general framework that may be taken into consideration for networking. A primary goal is to be engaged. Effective engagement requires that you believe you can make a difference and that others also believe that you can make a difference. The three significant components toward being able to do so require (1) trust between network members, (2) sense of belonging within a network, and (3) volunteering time and expertise. If this is successful, then the overall experience should lead to a net worth for you and you will feel that this network relationship is of benefit to you.

In my early years as a faculty member I learned of the importance of strategic plans for organizations. I immediately decided that I needed to establish a strategic plan for myself. If the success of an organization could be influenced by a strategic plan, then it was appropriate for me to establish a strategic plan to help shape the direction for my career. I found that having a strategic plan helped me conduct a SWOT (strength, weakness, opportunities and threats) analysis. Knowing what I needed early on was critical to identifying what type of networks and relationships I needed to build.

I now encourage any graduate student, postdoctoral fellows, assistant professors, or anyone starting a new career path, to start by building a strategic plan that will shape the direction they wish to embark on in the next 3-5 years. Establish your goals and have a metric and a timeline to let you know how you will know when you have met these goals.

Ask yourself:

Figure 1. Framework for effective engagement in a social network. Engagement requires that you believe that your contributions can make a difference, and that others also believe that you can make a difference. This requires trust, feeling a sense of belonging and having the required expertise. When this is accomplished the overall relationship will have a positive impact on the networking relationship.

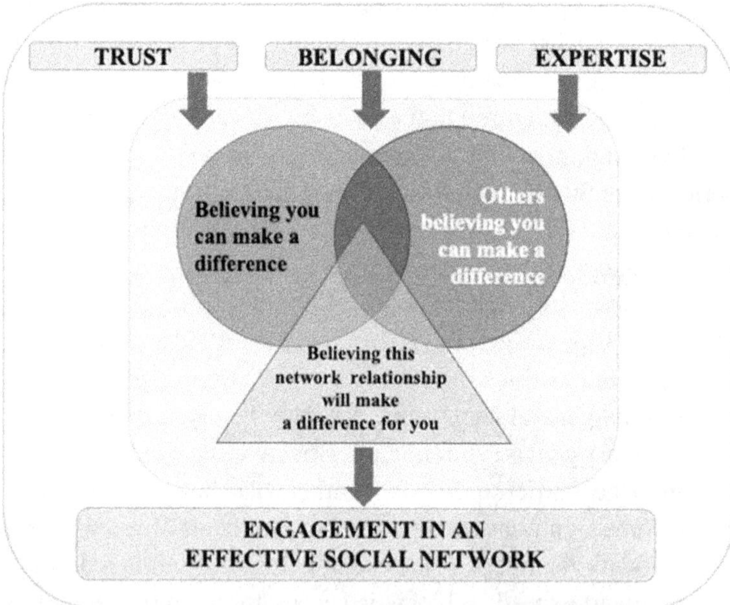

- What do you want to be known for and what type of network can help you get there?
- What type of network is valuable to you and what do you hope to gain from this network?
- Do you understand the culture of the environment you are in and how to succeed in it?

The network you build does not have to happen immediately, but it needs to happen strategically. It needs to be operational and transactional. Identify what your strengths are as well as your weaknesses. If you know your weaknesses, what type of networks will you need to build to compensate for that weakness. Identify where the opportunities are how you plan to take advantage of them. What type of networks will allow open doors for those opportunities? What are the potential threats, for example, others are there others in your field that will be competing for the same type of grants you are, and if so, how do you distinguish yourself and your work?

In addition to building your strategic plan, protect your time and make time to network. Whether it is with colleagues in your department, other units on your campus, community leaders or business leaders, make time to get to know them and have them get to know you. It is important that you spend time with others in informal settings so they can get to see who you are, learn about the great work you do, and see the value you bring to your organization or the discipline. Showcase your work and your accomplishments on your website. Keep it updated regularly because as you network people will try to read more about you. The easier you make it for them to find information that is comprehensive and in one location, the easier it will be for them to learn what you want to be known for.

Personal Experiences

I received a Doctor of Philosophy degree in Chemistry in 2002 from the University of South Carolina, Columbia, South Carolina. I then moved to pursue postdoctoral research at the Johns Hopkins University in Baltimore, MD from 2002-2004. Following a postdoctoral fellowship, I started my career as an assistant professor of Chemistry at Western Michigan University. I was the only person of color in my department and found myself quite isolated. I had been given excellent advice as a graduate student and as a postdoc, that I needed to write good proposals and get funding to my university. One of my strongest sources of support was through an interaction I had with a professor at Michigan State University. We ran into each other on a shuttle bus at the ACS national meeting. He too was the only person of color in his department and was eager to help an assistant professor just starting out her career. He took me to the NSF workshop at the conference and introduced me one program officer who introduced me to another program officer. Those relationships were critical to my career. Getting to know NSF program officers is extremely important, both those that are permanent as well as those that are permanent. At the time, I was not prepared to meet with NSF program officers, so I did what I knew to do best which is listen. One of the program officers gave me his card and asked me to send him my curriculum vitae. He indicated that NSF usually looks for scientists to serve on panels and if my background was a good fit, he would reach out to me. I was excited to learn that I had the potential to benefit NSF in some way, but I was more excited to have the opportunity to serve as a reviewer and learn more about the review process.

Several weeks later I was on my way to the NSF to serve as a panel reviewer. The experience I had as a panelist was priceless. I learned so much about what the nuts and bolts of what reviewers look for and that helped me become a better proposal writer. I got to understand what NSF reviewers' value which helped me learn how to do better quality of work.

I was able to secure more funding that enabled me to earn the funding that helped me get more articles published and soon I earned early tenure and promotion in my department. Right after tenure I found myself being pulled into more service in my department. I was asked to serve as the Director of Graduate Studies (DGS). While I realized it was a heavy lift, it was a role that mattered to me because I wanted graduate education in my department to succeed and I knew I could make a difference. At the time, what I did not know is that this role would teach me a tremendous amount about university administration. It was interesting how in my role as DGS I began to understand the roles of various offices at the university. It was this understanding that actually strengthened me as a faculty member and began to expose me to various institution leaders. I began to learn about the decision makers on campus, and the processes by which policies are instituted. It showed me more ways to be effective and extended my network. This experience as DGS gave me more confidence. As I spent more time outside my university at conferences I was better versed in my organization and in what leaders' value. My networking skills focused on what I can bring to the table rather than what I can take away. It was critical for me to understand my value proposition and learn how to articulate that in ways that would showcase me as a contributor. I found myself being invited to decision making places within the university, within the community, and within my disciplinary organization.

It was not simply about how to demonstrate myself as a contributor, but it also required that I did good work. It was important for me to be thoughtful in how I approached my scientific research work and how I it. Preparation is key! Understanding what information and what work is valued by various audiences, enabled me to prepare my work accordingly thus making me a better presenter. I have learned to do a lot of listening and observing. I have learned to be careful who I associate myself with. I have also learned to lend a helping hand when I can.

In 2014 I earned early promotion to full professor. A few months later I got a call asking me to serve in an administrative role as Associate Dean. Less than two years later I was asked to serve as Associate Vice President for Research. I served in that role for 25 months and then moved institutions to serve as a Dean. Below I provide various examples on how networking helped me overcome various barriers.

STRATGIES FOR EFFECTIVE NETWORKING

Creating Your Brand

Networking is considered an investment in social relationships, that have an important return on investment (Wolff et al. 2008). Networking is critical for career success

(Collins & Clark 2003). It is important for you to be viewed as a scholar who contributes new knowledge to the discipline, that also impacts society. You will find, as a woman of color in STEM, that you are constantly being pulled in various directions that can distract you from effectively pursuing your scholarly. Some of these activities include mentoring students, especially underrepresented minority students, being on various committees, community service. It is always important to be engaged in all these activities, but the time needed to build your scholarly work is essential to your career and to creating your brand (Beals 2011). More importantly, it is critical to those that will require your help and support. Always find ways to create opportunities for others to see the scholarly contributions that you have done.

Universities and colleges have marketing offices and it is imperative to get to know the staff who work there well. When you write research articles, win awards, or have grants funded, you want to make sure that the marketing office is aware, and that stories are written about you that accurately showcase your accomplishments. Be as involved as you can in writing the story to ensure that it is written is ways that are understandable to the public yet reflect accuracy and novelty. The story will need to be written in a style that is appealing to a broad audience, and your marketing office can help write it. These stories are critical and will be the foundation of what your networking circle will see. It is through these stories that they will be able to learn more about your value and your contributions.

Taking advantage of on-campus resources to market yourself:

1. What opportunities are available already on your campus? There is no need to reinvent the wheel. Meet your department chair and find out what resources are available to you, especially with respect to marketing.
2. Does your college have an associate dean for research? If so, what is their role in advancing faculty careers? Meet with them and share your CV. Ask them what they do and see if there is room for you to contribute. Tell them what you do and offer to provide service only if you can see this as bringing value to your scholarly work.
3. Once you have identified what is happening at the department level, and the college level, ask what is happening at the university level. Who is your program officer on campus? What services does the Office of Research (these offices have different names depending on what institution you are in, for example, Office of Research and Innovation; Office of Research Engagement; Office of the Vice President for Research) offer.

Where to Go to Network to Create A Brand

- Attend conferences in your discipline

- Volunteer your time effectively
- Organize a symposium at your disciplinary organization
- Invite speakers to your department
- Read announcements carefully (including those from offices of research at neighboring institutions)
- Participate on review panels
- Get to know the program chair of the symposia held in your discipline, and volunteer to chair a session, or anything they may need help with.

Building Your Network

As I began to build my network, as a woman of color I had to appreciate the skills of emotional intelligence. There is a lot that one can learn in terms of how one is perceived and treated and how you react to that. I always encourage everyone to take some time and read about emotional intelligence (Goyette, 2019; Steinbrecher et al. 2019) and these skills are critical to how you will be treated and how you handle the good interactions as well as the difficult ones.

The Networking Buckets

There are various areas of networking that all STEM Women of Color need to navigate. You can arrange these areas into two buckets. The first bucket is directly related to the expectations that your institution expects you to excel in. The second bucket is the bucket that often is not discussed but plays a central role in your career advancement. Table 1. focuses on areas of network related to your first bucket. While this is not a comprehensive list of all the networking sectors, it does provide some of the more critical areas that are important to add to your list.

Establishing your research network is the most critical component of your career. It will open doors for you at your current institution as well as any future institutions. Many of you have heard the term "money talks" – this is central to academic success. You have to have the funding to mobilize and accelerate your research projects. Connecting with program officers was critical. They had me serve as a reviewer and being amongst other reviewers helped me learn more about how funding decisions are made. It made me a better thinker and more critical of my own research. This made my approach to writing articles more critical and help me get manuscripts into quality journals in my discipline.

Understanding the leadership structure within your university is critical. During the first 5 years of my tenure at the University I did not understand the administrative structure, nor did I realize how important it was. I was focused on my research, my classroom and only interacted as needed with the Office of Research, the library,

Table 1. Critical relationships essential for establishing an effective social network

Research	Teaching	Service
· Journal Editors and Associate Editors · Potential collaborators · Program/Session/ Symposium/conference organizers · Leaders within your disciplinary organization · Program Officers at Federal Agencies · Researchers at National Laboratories · Instrument technical staff and other institutions · Center and Institute Directors relevant research	· Leaders in science education · Colleagues at other universities who teach courses in your discipline and/or subdiscipline · Textbook sales representatives · Textbook publishers · Textbook authors within your discipline	· High school teachers · Leaders in community organizations · Leaders in your disciplinary organization or related national organizations · Your local section organization

and the Dean's Office. I remember as an assistant professor working on a NIH educational grant that required institutional data. I had no clue that there was an office on campus whose job was to collect and analyze institutional data – The Office of Institutional Research. Had I known that it would have made my life so much easier. As I started doing more work within the university I learned of the various centers and institutes, and the various offices responsible for advancing the university. I also learned of the various committee structures and how decisions are made. This became extremely critical in my role as DGS. It meant that I could do a better job advocating for graduate students within my department. I was later offered the opportunity to serve on a committee focused on graduate education. I was exposed to national organizations that focus on strategies and best practices to improve graduate education. All this came as a result of being a contributor through networking on my own campus. The key is to listen and to identify what I could do to become helpful, and while helping organizations to identify how those can help me.

Get to know as many people on your campus who hold positions shown in Table 2. Get to understand what their offices do and learn what benefit they bring to the university. Learn how these benefits can add to your value proposition because you are as much an individual as much as you are a representative of the organization you work for. Therefore, their success is yours as much as yours is theirs.

As you get to learn about the organizational structures within your University try to learn their function even if it is not closely related to your discipline. The reason this is critical is in order for you to do good work in your research, or to write good proposals, understanding the organization as a whole and what you may or may not have access to, will enable you to determine what you can accomplish.

Articulating what you can and cannot accomplish as a result of you understanding your university is critical to how reviewers will evaluate your proposal. Often times the value that you bring externally relies of the strength of the university you work in. The various types of research centers and institutes and what they do may be a benefit to the department, your college and more importantly within your university. There are organizational structures within these structures. For example, what does your college's organizational structure look like. Who reports to who? Who makes decisions and how are decisions made? Are there committees that make decisions or is there a specific person within the organizational structure who ultimately makes the final ruling? What do they value? Centers and Institutes play a significant role in mobilizing research within a university. They are often started Who is in charge of various institutes and centers? Do not rush to be part of a center of institute simply because of its name or its appearance. Make sure the center has values that are aligned with your values. Make sure that it will benefit you and advance your career.

Table 2. On-campus Network Buckets

On-campus Network
· Chancellor/President · Provost · Deans and Associate Deans · Associate Provosts · Vice Chancellors/Presidents · Associate Vice Chancellors/Presidents · Representatives from the Office of Advancement/Development · Program Officers in the Office of Sponsored Research · Staff from the Office of Institutional Research · Representatives from Student Affairs · Faculty senate executive board · Human Resources · Office of the Registrar · The General Counsel

Before you approach any of these on-campus units, make sure you study the website carefully before you meet with them. Also, be clear on what it is that you are looking for. Examples include:

- How often do you get notification regarding the various proposals?
- How is limited submission handled?
- Does the university offer support for writing grants or reviewing grants? What is the timeline needed from the date of submission in order for you to get access to these services?

- Are there any workshops that the university offers to help in writing competitive proposals?
- Who are the successful faculty on campus who have had successful proposals, and can they introduce you to them?
- Do they have access to proposals that are successful that you can get access to?
- Is there funding to allow you to travel to visit program officers at national/ federal agencies?

In addition to knowing what the offices and roles do, make sure you get to know the staff. If it is an office you need, for example the office of research, build a strong relationship with your research officer. They will be willing to help you when you need it and when you least expect it.

In my role as an administrator I was able to see things with a slightly different view. When I became an associate dean for research in the College of Arts and Sciences at Western Michigan University, I oversaw committees and it was at that time that I began to understand how administrators and administration view faculty. Faculty who demonstrated success were those that always rose to the top. The stars were well known. They were the ones who would quickly tell you about their most recent publications, or their most recent grants. They were the ones who would visit and try to get to know their administrators well so they could ask for favors.

But there were a group of faculty members that did not do that, whose scholarly accomplishments were quite significant. They worked quietly and did good work, but they never told anyone of the great work they did. I called those our hidden gems. I wondered what it was that made the difference. It was how they communicated their success relative to others. Those whose success was communicated often were those that were well connected with their campus leadership. The great work of the hidden gems was never showcased. It is important to find ways to tell your story. Your university must celebrate your successes and it is your responsibility to figure out how to do that. Your college should have marketing specialists. Your university should have a marketing specialist. But one of the reasons these stories are not told is the difficulty in communicating STEM work especially fundamental science work in ways that are exciting.

Understanding the Organizational Structure of Your Disciplinary Organization

When I was a graduate student I was encouraged to attend and present my research at the American Chemical Society (ACS). I remember my first meeting at the ACS. It was held in San Francisco, CA. It had over 15,000 attendees. It was overwhelmingly exciting to be around so many people in my field at a given time, but I was also fascinated by how it all came together. I attended other conferences that were much smaller, but I always felt that there was something fascinating about what makes ACS the ACS.

Social Barriers

When I was a post-doctoral fellow, I attended a conference with a fellow male colleague. We both made it a point to attend a talk given by a researcher who was well respected in our field. After the talk we both had questions. We walked up to the speaker and my male colleague introduced himself and asked the question. They started having a dialogue and as they were having a conversation, I remember hearing the speaker say 'yeah, let's get together for a beer and chat more about your project'. I remember thinking to myself that they will connect over a beer and how simple they were able to network. They were not only about to have a networking session, they were going to have a bonding session, all over beer. I could never imagine myself getting together with a total stranger over beer. I could never imagine seeing myself being invited to go have beer.

Fast-forward, a few years later, as an assistant professor, I remember hearing my male colleagues on a Friday afternoon agree to have a get-together at 5:01 PM at the bar to have beer. I was too new to invite myself. It bothered me but soon came to the realization that this is how most gentlemen bond – over beer. That is how they network. Now I don't drink beer and I just don't like beer. But I became obsessed by how networking and bonding happens over beer. They not only get together to talk over beer, but they even have beer-brewing get-togethers. I realized quickly that I had to get smart about how I was going to engage in the networking world. I had a few challenges to overcome – as a woman, as a person of color, and as a non-beer drinker. I knew that networking was critical. It was what was going to make me or break me. And because beer and I were not going to mingle, I had to identify alternative ways of effective networking.

As a chemist I began to step outside my comfort zone and read books published by psychologists. I encourage all women in STEM to learn from the psychology literature and to also try to read books written for entrepreneurs. As a Woman in STEM you are an entrepreneur building your research enterprise, and you will find several books that will change the way you network, including the citations in this chapter. One of my favorites is "Influence without Authority" (Cohen & Bradford, 2017). In this book, the authors describe the way young people often view themselves as not having much to contribute because they may not have the right connections, or the resources to be viewed as being valuable. I knew that there had to be strategies that would have to work. I was also short on mentors and had no idea who to go to. I began to document every interaction I had with various colleagues, university administrators, faculty in other departments, staff, and students. How I interacted with seminar speakers as well as guest speakers to the university helped me learn more about networking and how to make the process work for me in ways that would be beneficial. When I went to conferences, I made it a point to be an observer. I watched how people interact with each other. I learned about body language, what to say and what not to say and when to say what and when not to. I kept reading more books about social interactions and what I learned is that there is a science to networking when you are a person of color.

Like most academics I suffer from the challenge of being an introvert. This coupled with the fact that I don't drink or like beer, does not help me in the networking arena. But what I learned fast is that you have to be viewed as having and adding value to whatever network you aim to establish. Value comes in all sorts of shapes and forms. In order to make progress you must understand what value you bring, and what value is appreciated by the person(s) you want to network with. And the two must be aligned. When they don't align you must detect that and make a decision on either to disengage or to correct it.

Beware of Who Want You in Their Network, I.E. Don't Get Used

I recall being in my first semester as an assistant professor and being approached by a male colleague who said that he "needed" me to serve as a co-PI on a NSF grant that he was going to be re-submitting. Initially I was excited to be invited to be on a NSF grant. Later that evening I pondered on why he said he "needed" me on the grant. I decided to have a dialogue with him on the project and how he would like me to contribute to it. It turns out that he had not taken the time to understand what type of research I do. This bothered me because clearly if he did not understand what I do, then why would he need me on a proposal. As I engaged with a longer dialogue with this colleague about the proposal, he shared that reviewers were concerned about the lack of diversity in the proposal and that as a "younger" woman and a "person of color", I would fill that gap. I decided at that moment I had a decision one of two decisions: I was going to have to either (i) explain to him that he could not just add me as a co-PI with no scientific contributions to the project otherwise it would not justify my role as a co-PI, or (ii) that it may be best to disengage from working with that colleague as he was clearly intending to mislead reviewers of my role. I battled with this decision for a few days. I asked myself if this is the type of network, I wanted to be part of. I wondered whether this partnership was valuable. After much thought, and despite the fact that this colleague had probably 20 years of experience in higher education compared to my 3 months, I decided it was time to educate him about what values a woman and a person of color could bring to the proposal beyond race and gender. I shared with him my publications, and I shared some of the awards I had received. I also explained that base on the work I do I did not see how this would align with his project and that I have never published any work on diversity, so he needed to find alternative ways to address diversity gaps in his proposal. I felt really good about my decision not to collaborate with him because I decided that this was not the network I need. I wanted to network with collaborators who had integrity. Integrity to me was not adding a person on your grant just because of their race and/or gender. That, to me, was misleading to a federal agency.

That experience with this colleague was a great learning opportunity for me. It taught me that often times people may wish to network with me just to claim that they have a person of color or a woman on their team. And I knew I was more than just a color and a gender. If people could not see my value beyond a race or gender, then that was not the network I needed.

I also knew that because of the small number of women of color in my discipline, I had to work on educating people that they need to understand what we have to offer in terms of intellect and scientific value. I knew I was going to be fighting an uphill battle and to me that battle was worth fighting even if my network would remain small.

STRATEGIES TO SHOWCASE YOUR VALUE

In this section, I share my personal experiences strategies I utilized that enable me to identify how to best showcase my value. I hope this narrative will be useful to readers in finding ways to demonstrate their value. In my early career, various conversations and advise about networking led me to understand that you network with the question in mind: What can I get out of this networking opportunity?

My question as I started my journey and even now, is whether this is really what we want out of networking. I believe the answer is Yes! However, what is the proper approach to establish a strong effective network. What if, you can plan for networking by starting with the thought of: How can I assist the person I am going to network with?

Let's think about this briefly:

If I seated you at a table next to the Editor of the journal that is highly valued in your discipline, what would that conversation entail? This is a question I posed to group of women about to embark of their academic career. Answers ranged from "what does it take for my paper to be accepted in your journal", to "Who are the reviewers", to "how do I make sure that you will not send my paper to be reviewed by some of the most difficult people to get my paper accepted. As an associate editor, these are questions I encounter all the time and frankly, they are questions that indicate that someone is trying to find workarounds. The best conversation I have had has been someone offering to serve as a reviewer. This individual asked "I enjoy reading articles in that journal and if you ever need an extra reviewer feel free to reach me. I will send you a copy of my resume and I would be happy to assist if you feel my background is valuable."

Here is what you just did:

- You created an opportunity to follow up with the Editor.
- Not only are you following up, but you will be sharing your CV which opens up a large number of opportunities (only do this if you have a nice CV to share)
- If the Editor sends you articles to review, this is an additional item that you can use for tenure and promotion.

When an Editor asks you to do a review, take this very seriously. Work hard on it. Be helpful to the editor. Being a reviewer takes time, but it also helps you demonstrate your knowledge. Editors appreciate reviewers who take their work seriously and you will stand out as not only a great reviewer but an excellent scientist. You will likely get more requests for the same editor and as you do consider strengthening that relationship, for example, invite them to give a talk at your university. They will appreciate that and soon you will be a strong part of their network. Eventually they will invite you to become part of other activities they are engaged in – this is the remarkable model of reciprocity. This type of volunteering requires trust from both the reviewer and the editor, and part of building a network is being able to build that trust, Trust is built over time, through timely responses to requests, and by doing a good job when called upon (Band 2019; Schreiter et al. 2015).

The difference between scenario I and scenario II is that in scenario I you took a list of do's and don'ts and now you are going to go home and try to apply them on your own. In scenario II you created a relationship with the Editor so that you can have a back-and-forth relationship. In a few months or years, as a valued contributor, you will be able to get so much more out of that relationship, than simply taking some dos and don'ts.

Always ask yourself the question: "*What can I bring to the table*? Because, you see, as a person of color you are likely to be perceived as needing something, wanting help, needing support, and yes you do. But you also have a lot to offer. Find out what it is that you have to offer. Because when you are seen as a contributor and not a consumer, you are viewed differently. You now come with value, and it is that value that you bring that will get you a seat at the table. Always think about the value you carry and how can you bring that to a networking opportunity so that you leave something memorable with those you come in contact with.

Learning to Network Within Your University and Outside Your Department

It was a Sunday morning in February 2015 when an email arrived from an administrator asking if I would consider serving in a leadership capacity. The time had come for me to step over to the dark side of academia. And frankly, I was ready for new challenges. I was ready to work in ways to help make my institution better. I knew that the reason I was being approached about this opportunity was because someone had considered me to have some type of value to bring to university leadership. The position was Interim Associate dean for the College of Arts and Sciences. I was honored to be invited and considered for this role. I also realized that this was going to be one huge learning curve. The college had 26 departments and I did not know may people. I realized that I was going to have to find a different way of networking. Unlike what I had learned from the "*Influence without Authority*" book, for the first time I was actually in a position with some type of authority over those that I would be working with. It was going to be an interesting ride with regard to networking. What I knew is that I had a lot to learn in order to be an effective leader. But I wondered who the right people would be to offer me the support I needed to learn what I needed to know. I also did not know what I did not know, which meant I needed to network in a very different way.

I remembered reading a phrase that said, "you were given two ears and one mouth because you need to listen more than you should talk". That resonated more and more with me in my newly acquired role. I had to do a lot of listening to better understand several things, ranging from what are the true challenges that the college was facing, what are the various departmental challenges, who do you go to, and who do you trust. There is so much to appreciate in a discipline diverse college such as Arts and Sciences. One of the duties I was assigned was to serve as Acting Director for the Africana Studies program. I was also charged with all aspects of diversity of inclusion. I admit I have never taken any courses in Africana studies, nor have I had any formal training in diversity and inclusion aspects. But clearly, I was assigned these responsibilities because of my race. And when you are asked to serve in a role to support such initiatives you are obligated to figure out how to do the best job you can to be supportive. I leveraged all the resources I could. I reached out to the Vice President for Diversity and Inclusion. She had tons of resources and she encouraged me to meet the various leaders in the university. But before I met each of them, I made sure I knew what they did, what they valued and what my contributions should and could be. That is exactly what I did. I studied university administration. I read the literature and made sure I understood what each office did. I listened a lot. And soon I was engaging in conversations and taking notes on what they needed to be done. They also became the people I would call on in order to get things done. I learned so much about higher education and it was all because I built a strong network. And when people in my college came to me with problems, I knew exactly who to call and what questions to ask. In university administration you have to be viewed as a problem solver – that is your value and the asset you bring to the table. You have to respond to others, and you need to respond quickly. You must be seen as someone who follows policy and procedure, who is fair and has integrity. And when people begin to understand that you have integrity and that you follow process, they will respect that because they understand how decisions are made. You have to be authentic. Do not try to be a person you are not. I have stood firm in my beliefs which are to always follow policy and procedure, treat people with utmost respect, to treat people fairly and equally, and never pretend to be someone I am not. I have forged respectable relationships that I will treasure throughout my life.

CONCLUSION

Networking happens at various levels for numerous reasons. Always be prepared to build your network by being able to share who you are and what values you bring to the table. It is always important that you take a listening role when meeting new people and really hearing what they value so you are able to present your unique qualities in ways that they see this as valuable to what they do. Take the time to decide what you would like to be recognized for and how you want people to value you. Identify your strengths and your assets. If you are not sure what they are, consider making a list of values and assets you aim to have in the future and work your way to establish them. Your network is one that will evolve over the years and as you build relationships the network will grow. Accept that there will be relationships you will want to build that may sometimes work, but in some cases will not. The academic world is small, and you will find that people in your discipline know each other, therefore building your brand within that community will serve you well. And as you build your networks you will find that you are being called upon often. Just remember that it is alright to say no when you know you just can't accommodate the request being made. When you do say yes make sure you are answering the call in a timely manner and do it well. Do it so well that they will come back to you for more. This is how you establish what you want to be known for. For me, I hope to be known as someone who gets this done! It is that simple – just getting things done.

REFERENCES

Band, Z. (2019, May 15). *Success Is in Your Sphere: Leverage the Power of Relationships to Achieve Your Business Goals*. McGraw-Hill Education.

Beals, J. (2011). *Self-Marketing Power: Branding Yourself as a Business of One* (1st ed.). Keynote Publishing LLC.

Bradford, D. (2014). *Up Your Game: 6 Timeless Principles for Networking Your Way to the Top Life*. Science Publishers.

Coburn, D. (2014). *Networking Is Not Working: Stop Collecting Business Cards and Start Making Meaningful Connections*. Influential Marketing Group Publishers.

Cohen, A. R., & Bradford, D. L. (2017). *Influence without Authority* (3rd ed.). Wiley Publishers.

Collins, C. J., & Clark, K. D. (2003). Strategic human resource practices, top management team social networks, and firm performance: The role of human resource practices in creating organizational competitive advantage. *Academy of Management Journal, 46*(6), 740–751.

de Janasz, S. C., & Forret, M. L. (2008). Learning the Art of Networking: A Critical Skill for Enhancing Social Capital and Career Success. *Journal of Management Education, 32*(5), 629–650. doi:10.1177/1052562907307637

Fisher, D., & Vilas, S. (2000). *Power Networking: 59 Secrets for Personal & Professional Success* (2nd ed.). Bard Press.

Goyette, K. (2019, July 30). *The non-obvious guide to Emotional Intelligence (You can actually use)*. Ideapress Publishing.

Lippe-McGraw, J. (2019, April 16). *Simple Ways to Network, Even If You're Shy, It goes far beyond LinkedIn*. https://www.oprahmag.com/life/work-money/a27132577/how-to-network-effectively/

Marcus, B. (2019, December 3). *The Networking Advice No One Tells You*. https://www.forbes.com/sites/bonniemarcus/2018/05/22/the-networking-advice-no-one-tells-you/#34a2402d7772

Schreiter, K., & Schreiter, T. (2015). *How to Follow up with you marketing Network Marketing Prospects*. Houston, TX: Fortune Network Publishing.

Spurk, D., Kauffeld, S., Barthauer, L., & Heinemann, N. (2015). Fostering networking behavior, career planning and optimism, and subjective career success: An intervention study. *Journal of Vocational Behavior, 87*, 134–144. doi:10.1016/j.jvb.2014.12.007

Steinbrecher, S., Schaefer, R., & Moyle, J. (2019). *Meaningful Alignment: Mastering Emotionally Intelligent Interactions at Work and in Life*. Sustainable Path Publishing.

Wolff, H. G., Moser, K., & Grau, A. (2008). Networking: Theoretical foundations and construct validity. In J. Deller (Ed.), *Readings in Applied Organizational Behavior from the Lüneburg Symposium* (pp. 101-118). Mehring, Germany: Rainer Hampp.

Chapter 5
Navigating the STEM Professoriate as a Black Woman:
Surviving but at What Cost?

Nichole L. Powell
Emory University, USA

ABSTRACT

The absence of faculty diversity continues to plague American colleges and universities, with about 6% of associate and 4% of full professors identifying as Black or African American, based on a 2016 survey conducted by the National Center for Education Statistics. This chapter will reflect on the personal experiences of a Black woman in STEM, examining whether diversity efforts have affected the number of Black and African American women in the STEM professoriate. It includes current efforts regarding diversity and inclusion, as well as recommendations of ways that institutions can help to foster a sense of belonging.

INTRODUCTION

The lack of diversity in institutions of higher education is well documented, yet despite calls to increase faculty diversity in the academy, the percentage of Black faculty has remained essentially constant. The number of articles regarding the experiences of students, staff, and faculty from underrepresented racial and ethnic groups, point to the continued issues regarding institutional climate. This chapter discusses definitions of diversity and institutional reporting of diversity gains. The account

DOI: 10.4018/978-1-7998-4858-5.ch005

uses an autoethnographic approach and includes the reflections of a Black woman in STEM with experience working at both a historical Black university (HBCU) and a liberal arts intensive predominantly White institution (PWI); examining whether diversity efforts have affected the number of female Black and African-American women in the STEM professoriate. The chapter ends with current efforts regarding diversity and inclusion, as well as recommendations of ways that institutions can help to foster a sense of belonging.

BACKGROUND

Definition of Diversity and How Institutions Define Diversity Gains

Diversity may be described in various ways. The D5 coalition, one of many groups committed to advancing diversity, equity, and inclusion, defines diversity as a "demographic mix of a specific collection of people" focusing particularly on racial and ethnic groups, LGBT, persons with disabilities, and women ("What is DEI?, " n.d.). Their definition "encompasses what diversity looks like in the 21st century" and focuses on groups that have remained underrepresented and "marginalized in the broader society." They define equity as "impartiality and fairness within the procedures, processes, and distribution of resources by institutions or systems." Stating that "this requires an understanding of the underlying or root causes of outcome disparities within our society." Most institutional efforts focus on structural diversity - which refers to the "numerical representation of various racial/ethnic groups" (Hurtado, Milem, Clayton-Pedersen & Allen, 1998).

With various campus climate surveys and assessments, how do institutions decide which programs to develop or what changes to make to increase the recruitment and retention of minority faculty? With such few minority faculty, and the ability to be easily identified, how willing are minority faculty to give honest input? How reliable are the results obtained from such surveys? Vaccaro (2010) showed how positive campus climate surveys can hide serious issues related to institutional sexism and racism. A recent article on the results of a campus climate survey conducted at the University of Iowa and the University of Maryland - College Park stated that minority faculty at both institutions "expressed lower levels of satisfaction compared to their majority counterparts and cited both overt and subtle forms of bias" ("Campus Climate Survey Findings Break Along Color Line," 2019).

The American Council on Education's Annual Status Report of Minorities in Higher Education Fall 2016 gives the percentage of Black full, associate, and assistant professors as 3.8%, 5.8%, and 6.3%, respectively (Snyder, de Brey, &

Dillow, 2016). These numbers are the same for NCES faculty data reported for surveys conducted in 2009, 2011, and 2013. The percentage of Black female full, associate, and assistant professors is about 2%, 3%, and 4%, respectively. The overall percentage of minority faculty at the full, associate, and assistant professor levels are 23%, 21%, and 16%, respectively, showing that since 2009 there has been a 2% increase in the number of full professors and 3% in that of associate professors. These data indicate relatively no change in the amount of Black faculty despite various institutional diversity initiatives.

A 2017 survey by CUPA-HR quantifies issues faced by minority women in terms of representation and pay with women of color and men of color earning $0.67, and $0.72, respectively, per $1 earned by White men, while White women earned $0.81(McChesney, 2018). The report indicates that men of color are underrepresented in all areas of higher education when compared to the general and the college-educated populations. In contrast, White men are underrepresented only in staff and professional positions. Of all the groups, women of color show the largest underrepresentation in faculty and administrative positions. White women were not underrepresented in any of the assessed categories. While institutions focus on overall structural diversity, these studies indicate the need for more attention and policies regarding racial and ethnic diversity, acknowledging that these groups have faced systemic disadvantages, with an understanding that equity means that resources need to be allocated to account for these disparities. Without equity, along with an inclusive campus and higher education culture, it is difficult to cultivate the sense of belonging that will result in significant changes in the numbers of tenured Black faculty.

A BLACK FEMALE STEM FACULTY MEMBER'S JOURNEY

I grew up in the Caribbean, where I experienced classism and some colorism but not overt racism. I was taught that education was the way to traverse social classes and improve one's social status. I gave little attention to the impact of race since the country's decision-makers were of various shades, albeit many were of a slightly lighter hue.

My journey in the United States began as an international graduate student at a public university in Georgia. While there were many international students in my cohort, I don't recall any of African or Caribbean descent. However, I did not feel isolated as the Ph.D. program had other African-American, Black, Caribbean, and Hispanic students. As I acclimatized to my new environment, I felt the effect of the color of my skin on my treatment in the general public - this was new and unfamiliar. I was watched and followed while shopping in stores, often being given "special" attention and asked if I needed help. White patrons were able to enter and browse without much notice. Somehow, I did not believe that such discriminatory practices were common-place in academia. I knew that there were few minority faculty members, but I was unaware of the disparity in the way they are treated. I entered the faculty ranks with a lack of anticipatory socialization; having little knowledge of the professoriate or higher education institutional policies and culture (Tierney & Rhoads, 1993). As a new Ph.D. in an adjunct faculty position, I realized that I was being treated differently. A particular faculty member always called me by my first name while addressing other professors as "Dr…". I often found my mail opened but thought it was a mistake. After observing the faculty member's interaction with me in class on a few occasions, one of my White female students called my attention to the fact that I was being treated differently because of my race. Her parents were professors at a PWI under scrutiny for racial discrimination among their faculty, and she opened my eyes as she recounted the problems faced by minority faculty at that institution.

My first tenure-track position was at a HBCU. Many students did not have Black high school teachers and welcomed being part of a community of Black intellectuals. Yes, the environment had its equity issues, namely gender equity, and in spite of them, I felt accepted and grew professionally. As a junior faculty member, my voice was encouraged; I made recommendations and helped to implement changes across the university. Though I lacked resources, I gained confidence. I am indebted to those who saw my potential and allowed me the freedom to grow. During this period, I had a senior faculty member as a mentor who encouraged my interest in curriculum development and academic affairs. I remember my mentor asking me what I really wanted to do. It was through that reflection that I realized that my institution was not able to offer what I needed. My time at that institution felt like the induced-fit model of enzyme catalysis: I did not have to, nor was I ever asked or expected to, conform to the mold to be a good fit.

Joining the Faculty Ranks at a PWI

I came to my current position fully cognizant of the lack of faculty diversity at my new institution. Motivated by the need to be an asset to my Black community, for students, particularly Black students attending a PWI to see that there are Black women in STEM, I was unprepared for the realities I faced. While I had spent several years on the tenure track at my previous institution, my experience did not translate to this new setting. Instead, it has been more like the lock and key model of enzyme catalysis: conform in order to fit.

I began my position with little understanding of what the requirements were for promotion and tenure. My department provided no help or guidance, and I quickly realized that a newly-hired faculty member from another science department received her own lab, fitted with all the new equipment necessary for her research. In contrast, no one asked me what I needed; I was just given some start-up funds. A junior colleague in my department pointed me to funding sources to obtain research equipment and helped me navigate that process. I loved working with two of my colleagues on curriculum ideas, apart from that, I felt unwanted. At one point, I asked, "Who didn't want me here? Was I the diversity pick?"

Did you want to see me broken?
Bowed head and lowered eyes?
Shoulders falling down like teardrops,
Weakened by my soulful cries?
Does my haughtiness offend you?
Don't you take it awful hard
'Cause I laugh like I've got gold mines
Diggin' in my own backyard.
You may shoot me with your words,
You may cut me with your eyes,
You may kill me with your hatefulness,
But still, like air, I'll rise
Maya Angelou (1978, p. 41-42)

I reflect on the institutional and cultural structures that challenged my success as well as systems that provided support. My experiences include gender and racial challenges, lack of mentoring and meaningful feedback, microaggressions, and microinvalidations, as well as challenges with work-life balance.

GENDER AND RACIAL CHALLENGES

Hostile Classrooms: Impact on Teaching Evaluation

In the classroom, a Black female faculty member is expected to be motherly and nurturing. In keeping with this, I have often found that students expect me to change exam dates, grading criteria, give special concessions, and are angry when I do not. Studies show that African American and Black women have to be careful of their presentation (physical, verbal, and non-verbal) in the classroom as these affect their evaluations (Merritt, 2008). Students have challenged my subject knowledge, but not in a way to promote learning. Instead, strategies include asking specific questions based on journal articles they went out of their way to read, doubting, questioning, or dismissing what I say based on what their extraordinary AP/IB teacher told them.

Many institutions use Student Evaluations of Teaching (SETs) as the primary means of evaluating teaching effectiveness and as an essential component of the tenure evaluation process. But what do these teaching evaluations measure? I realized issues in the use of such a SET within my first year at my current institution. My scores for what should be objective measures such as "returns assignments in a timely manner" were not reflective of what was being done and were lower than those of my colleagues. I consistently received a low 4 out of 5 overall rating, although all quizzes and exams were graded and returned to students during the next class session. So, what would it take to improve my score on that item? The prompt return of assignments was one example of the effect of gender bias on course evaluations mentioned by Boring and colleagues (2016) noting that "SETs were more sensitive to student's gender bias and grade expectations" and these biases can "large enough to cause more effective instructors to get lower SETs than less effective instructors." Other articles discuss the influence of race in SETs, acknowledging that such biases exist and some suggest ways of dealing with them (Nast, 1999; Steiner, Holley, Gerdes, & Campbell, 2006; Wallace, Lewis, & Allen, 2019).

I also realized the influence of the likeability of the instructor while testing a survey I developed for a Scholarship of Teaching and Learning project. A particular question that was very specific to activities done only in my class (e.g., related to the reading of a journal article) was scored highly by students in a colleague's class. The subjective answering of an otherwise objective question has made me skeptical of the use of any evaluative instrument that does not include the opportunity for open-ended responses. While the "test survey" elicited response by asking students to explain or give specifics regarding the question, only the students in my class wrote related comments/responses. One student in the other class who correctly stated they had not had that experience wrote: "Are you joking?" I could not have determined that students had not read a journal article in their class, without asking them to include specifics such as what, why, and how.

Research conducted by Carrel and West (2010) found that teachers who contributed most to student learning (i.e., deep learning, and course success) in subsequent courses had received lower student ratings in their prior course. Other factors that affect student evaluations include non-verbal behaviors, including personality and cultural factors. The need for good student evaluations for promotion can lead to issues related to identity as one may need to change to fit expectations rather than to be their authentic self. Several published articles draw attention to the effect of gender-related expectations on course evaluations, including discussion of the view of the Black female faculty as "Mammy" by students, faculty, and administrators (Smith, 1999; Sprague & Massoni, 2005). Smith (1999) describes Mammy as *a Black woman who acquiesced to her inferiority and White supremacy, her excessive love for, and protection of her young White charges. She epitomized the extent to which all Blacks should be willing to sacrifice.* She further states that Black women not fitting the role of Mammy can be seen as "authoritarian, uncaring, and adversarial" and are often penalized through "excessive negative evaluations."

Devalued Scholarship, Teaching, Service, or Contribution

The stated criteria for promotion and tenure is the successful demonstration of achievements in scholarship, teaching, and service, the three-prongs of the tenure process. In about my fourth year, I was informed that my laboratory-based project(s) were not making sufficient progress. Additionally, I needed to have a lab-based publication in my tenure package to show that I am a chemist. The tenure dossiers of previous science-based tenure recipients did not indicate a lab-based publication requirement. It was also communicated to me that I needed at least one publication before tenure. Post-tenure, I heard a change in this message in regards to tenure-track candidates in the pipeline - they need to show a trajectory that will result in publication. I am left bewildered by subjectively changing requirements and standards.

One colleague, having had several meetings with my research student (unknown to me), came to my office and stated that they knew what the issue was with my research that was impeding my progress. Apart from having secret meetings about my research, I could have taken this "insight" more positively if it was not a reflection of the lack of respect for my intellect and a presumption of my incompetence as a researcher and scholar. It was disheartening to realize that although I submitted annual reports, had a third-year pre-tenure review, and presented a successful tenure portfolio, my colleagues seem to know nothing about my graduate training, teaching philosophy and values, or my research interests, presentations, and publications. An example of the invisibility and invalidation I experienced was when my colleagues failed to acknowledge my area of training, stating that we did not have the expertise regarding a new course offering. Another experience was during discussions regarding the need to find resources that are related to my research interests and publications, my ability to contribute was overlooked and devalued, even after explicitly bringing attention to my knowledge. Being ignored or having my ideas attributed to someone else is more common than I would like it to be; however, I have found that other Black female faculty have had similar experiences (Settles, Buchanan, & Dotson, 2019).

The reality is that underrepresented groups are held to a higher standard. It is, therefore, vital that we ensure that we are meeting the expectations (as we understand them) necessary for a successful tenure dossier. We may find that our attempts to do what we deem as essential for our success may be met with a lack of support and criticism. Focusing on our scholarly productivity may be seen as not being a team player, or putting our interests before that of the department, division, or school. Is this the "Mammy" expectation at play? Would a White male ever be told that they seem to place their scholarly pursuits and professional aspirations above the needs of their department?

Service expectations vary from institution to institution, and what counts as service may even differ among the departments and schools within the same institution. However, when using equal "measures," what is difficult to rationalize is having seemingly different expectations from colleagues of other racial and ethnic groups. What counts as service, and how does the service focus differ based on race and/or gender? The article "Research, teaching, and service: Why shouldn't women's work count?" discusses the devaluation of teaching and service, areas of the tenure evaluation process that are most likely to be the focus of female faculty (Nast, 1999; Steiner, Holley, Gerdes, & Campbell, 2006).

Lack of Mentoring, Meaningful Feedback and Social Support

Earlier, I mentioned that my department provided no help or guidance in my transition to the institution and my understanding of the tenure and promotion requirements and process. Zambrana and colleagues (2015) detail the importance of mentorship, including political guidance - help in understanding "institutional norms, the role of race and relations" in the success of faculty from traditionally underrepresented racial groups. The feedback I received was primarily in the annual letters from my department/division and were not very helpful in guiding my growth as a teacher or scholar.

I also experienced a lack of social support. As the only Black female faculty member on campus, there was no community of minorities at the institution to lend support or feedback. It was also difficult when, in their attempt to be supportive, other women lump themselves with my experience while neglecting their privilege. At times, I feel like I have no voice. Even when explicitly given a voice, attempts are made to minimize it or take it away. Gilbert and Tatum (1999) state that social support is essential for the progress and well-being of Black female faculty.

Some of my positive experiences include my participation as an external committee member in a department/division faculty search. The committee was comprised solely of women who ensured that my voice was heard - often making sure to ask my opinion before voicing their own. They also expressed their reservations of giving too much weight to SETs - acknowledging their subjective nature.

Other senior White female faculty encouraged me to use open-ended questionnaires as an additional way of collecting teaching evaluations in order to provide me, and others, with a better picture of the students' classroom experience.

Microaggressions and Microinvalidations

Microaggressions and microinvalidations exist, and one of my jaw-dropping moments occurred during discussions regarding recruiting diverse faculty. The discussion included questions such as: "are you asking us to lower our standards?" and other sentiments related to "fit" such as: understanding our institution means that they attended a similar institution, and the need to hire from one of the top graduate programs in the discipline. Gendered microaggressions affect the work environment of female faculty, with a recent study showing that STEM female faculty of all career stages and ranks experience microaggressions (Yang & Carroll, 2018).

The treatment of students from underrepresented groups is also a telling sign of the racial climate: such as the lack of Black students in departmental roles, research positions, or receiving academic awards. I have heard comments such as "in all my years of teaching, he is the first Black student who did very well in my class."

Work-Life Balance and Challenges

The culture and policies of higher education institutions continue to utilize a reward system based on the traditional male model, forming systemic barriers to the success of female faculty (Bird, 2011). Knowledge of the resources available and having a supportive working environment is essential. Many women face changing family roles during their career: from their role as a caregiver of a child - be it a newborn, newly adopted or foster child, or the realities of dealing with illness or ongoing special needs. Often, Black female academics lack the extended family support structure that would allow them to balance their professional and personal roles easily, or they have to conform to and fulfill traditional cultural roles. At the other end of the spectrum is the role of being a caregiver to ill, special needs, elderly parents or other adult family members. It could be assumed that such responsibilities can be easily fulfilled through the use of facilities such as daycares and assisted living services, but that assumption fails to recognize the socioeconomic factors that impact such access. Data obtained from a 2003 faculty survey indicate that Black female faculty accounted for about 0.6% of the faculty households that earn $100,000+ per year compared to about 16% of White males, with this group accounting for the largest percentage of White male faculty households (Snyder et al., 2016). About 48% of all White male faculty households earned above $70,000. The homes of a majority of Black female faculty earned between $25,000 - $39,900 or $55,000 - $69,900, each accounting for about 21% of Black female faculty households. This disparity in household earnings is a socioeconomic factor that may cause Black female faculty additional stress. Additionally, many Black female faculty have to relocate for job opportunities and often lack readily available family support (Harley, 2008).

Additionally, institutional resources related to life changes are often not well advertised. I was not knowledgeable of available resources as I dealt with my child's health-related issues, including hospitalization for several months. I was not offered or informed of ways to reduce my work-load, teaching, and other obligations. Looking back, I have difficulty comprehending how I did it all. Despite searching the human resources (HR) documents and policies, I have not found guides regarding work-load and teaching obligation as they relate to caregiving responsibilities. FMLA policies are not advertised and are not easy to interpret. A recent article found that women are less likely to find tenure expectations as being reasonable, where reasonable expectations include family-friendly policies such as caring for an ill relative (Lisnic et al., 2019).

My Transformation

I survived the tenure process, and the support I received from institutional administrators was vital. They provided helpful advice to navigate the various situations I faced, and they have been my primary sponsors, demonstrating the institution's commitment to faculty diversity. Recommendations I received from them were mainly record-keeping in nature. Examples include ensuring that my senior colleagues saw my syllabus (with any scheduled missed/make-up classes) following being reprimanded for missing class to present my scholarship at a conference, despite advance notification. After being criticized for not being on campus (due to health reasons), I was advised to email my senior colleagues and copy the administrative assistants whenever I would not be on campus. I had to justify the scheduling of my office hours, and I posted my office hours and teaching schedule on my door after my annual letter stated that students and faculty were unable to find me. These types of record-keeping help but they do not prevent or acknowledge the possible psychological, physical, and other health-related damage (Harley, 2008; Keashly & Neuman, 2010) of such hypervigilance. Unfortunately, there seems to be little or no repercussions for the perpetrators and an overall lack of understanding by administrators that anti-retaliation policies provide little to no protection in academic institutions where racism is systemic. That discrimination can be justified using nebulous standards and phrases such as lack of fit, lack of productivity, and failure to meet promotion standards.

I recognize the mental, emotional, and other health-related scars that persist as a result of this journey. My recent participation in the National Center for Faculty Development and Diversity's (NCFDD) Faculty Success Program provided the critical reboot I needed. I had no idea these faculty development resources existed, and my participation was sponsored from my institution. Now, I am focused not only on surviving but also on thriving, giving appropriate priority to self-care, personal and family obligations, and achieving the work-life balance necessary to navigate my multiple identities and be my authentic self. I am a Black female STEM faculty, mother, sister, daughter, niece, aunt, partner, friend, caregiver, advisor, and mentor.

I think the importance of doing activist work is precisely because it allows you to give back and to consider yourself not as a single individual who may have achieved whatever but to be a part of an ongoing historical movement. Angela Davis (1997).

While this chapter details some of the challenges faced based on having multiple identities, it also includes the encouragement and support received while navigating the professoriate. It demonstrates the importance of having an institution and administrators who are deeply committed to diversity, equity, and inclusion. While self-confidence and self-reliance have been essential in this journey, professional gains are difficult without the appropriate support and sponsorship.

Now, on the other side of the tenure process, I find it important to stand in the gap to help faculty from underrepresented groups navigate the tenure process while assisting institutions to understand the issues that we still face. I see myself as a change agent, an activist, changing the way institutions see and value their employees as well as the resources available to support our well-being and work-life balance. It means proposing and supporting change integral to increasing the number of Black faculty - this includes changes related to encouraging and supporting the success of Black science students. The number of Black female faculty will not change until we change the institutional culture, and provide the necessary support and resources to address the underlying issues regarding transition to faculty careers and success, including having family-friendly policies that are encouraged and supported within departments and by administrators. These changes include making HR policies visible and transparent and making parental and family leave the norm and not the exception. Such support includes mandating flexibility in the tenure clock, allowing extended tenure cycles, all without detriment to tenure decisions.

RECOMMENDATIONS

Each faculty member from traditionally underrepresented groups should be seen as a unique individual, with unique attributes and needs. Below are a few recommendations to help support the success and sense of belonging with particular reference to Black female faculty.

- Provide access to mentors who understand the issues faced by faculty from underrepresented racial groups and commit to providing and supporting mentoring relationships

The article "Black female faculty success and early career professional development" discusses the experience of Black, female faculty, and the importance of having similar mentors. While having a Black male mentor is good, he lacks the "gendered experience" faced by female faculty (Jones & Osborne-Lampkin, 2013). Also, Tillman's (2001) article on mentoring African-American faculty highlights the professional and social isolation Black academics face, and the importance of the psychosocial support offered in same race mentoring relationships (Tillman, 2001). Institutions may find it difficult to pair new faculty with mentors of the same race due to a lack of critical mass, but it is crucial that mentors are provided who understand the issues faced by Black faculty at that institution in order to provide adequate guidance for navigating the process. Additionally, national and regional Black mentoring and networking resources should be made available without cost to or payment from the funds available to the faculty member.

- Provide access to resources needed for success

It is important to directly ask each faculty what resources or opportunities they believe would be helpful — with the understanding that many faculty may not even be fully aware of their needs. This lack of awareness often stems from a lack of understanding or knowledge of available resources as well as the historical perception of total self-reliance. Indeed, many faculty from traditionally underrepresented racial groups may have graduated from institutions where they did not have the resources or mentorship in areas necessary for their success as faculty member. Having faculty development resources such as grant writing training, assistance with understanding the components of a successful tenure portfolio, would be helpful for not just underrepresented faculty but for all faculty. Often faculty do not realize their needs until they explicitly see the resources advertised. Sometimes they are not sure of their needs until they are fully engaged in their project, therefore having a supportive, non-judgmental environment would be helpful.

- Facilitate intentional connections and community building opportunities with other racial and ethnic minority faculty across the institution

Intentionally provide venues and opportunities for faculty from underrepresented groups to have a non-judgmental space where they can discuss challenges, ideas, provide psychosocial support, and have the opportunity to learn from each other.

- Actively protect minority faculty

Once hired, minority faculty should be given the assurance that their department, school, and the institution is committed to their success. Such guarantees should be made in both words and actions. When minority faculty voice concerns or recount unfavorable experiences, these should not be ignored as being overly sensitive. People of color and women experience bullying in the higher education workplace at a greater level than in other work environments, and a much higher rate than Whites and males (Hollis, 2017). Additionally, while institutions often cite policies regarding retaliation, minority faculty understand that it is often difficult to "prove" such retaliation or bullying (Keashly & Neuman, 2010).

- Encourage the inclusion of service as Scholarship of Engagement

Black female faculty tend to engage in community service as a way of giving back to their community. Articles such as "Faculty service roles and the scholarship of engagement" discuss the value of giving greater recognition of these important, and often overlooked service roles (O'Meara, 2002; Park, 1996; Ward, 2003).

- Conduct exit interviews of minority faculty who leave their positions or those who decide not to accept a job at the institution

The recruiting and hiring of established Black faculty members (ie. tenured at other institutions) is useful for building critical mass, but should not be used as a measure of institutional health. Institutional climate should be measured by the ability of Black faculty to progress through the ranks from assistant to full professor, at a percentage and rate similar to that of White males.

Unfortunately, unfavorable promotion and tenure decisions can be accompanied by justifications that are often difficult to refute. It is therefore vital that departments and colleges be held responsible for giving adequate guidance, mentorship, and support, and that the provision of such assistance is assessed at least annually by the department, faculty, and the office of the Provost or Dean of Faculty. The success of the faculty member must become the responsibility of the entire administration of the institution.

Prejudice can be actuated by different reasons. Sometimes the reason is economic, and sometimes political. You can only obstruct it by progress and force. Marcus Garvey (1968 p. 16).

REFERENCES

Angelou, M. (1978). *And still I rise: A book of poems*. Random House Publishing Group.

Bird, S. R. (2011). Unsettling universities' incongruous, gendered bureaucratic structures: A case-study approach. *Gender, Work and Organization*, *18*(2), 202–230. doi:10.1111/j.1468-0432.2009.00510.x

Boring, A., Ottoboni, K., & Stark, P. (2016). Student evaluations of teaching (mostly) do not measure teaching effectiveness. *ScienceOpen Research*. doi:10.14293/S2199-1006.1.SOR-EDU.AETBZC.v1

Campus Climate Survey findings break along color line. (2019, October 17). Retrieved December 1, 2019, from *Diverse* https://diverseeducation.com/article/157754/

Carrell, S. E., & West, J. E. (2010). Does professor quality matter? Evidence from random assignment of students to professors. *Journal of Political Economy*, *118*(3), 409–432. doi:10.1086/653808

Davis, A. (1997). The two nations of Black America. *FRONTLINE | PBS*. Retrieved November 29, 2019, from https://www.pbs.org/wgbh/pages/frontline/shows/race/interviews/davis.html

Garvey, M., & Garvey, A. J. (1968). *Philosophy and opinions of Marcus Garvey*. New York: Arno Press.

Gilbert, E., & Tatum, B. L. (1999). African American women in the Criminal Justice academy: Characteristics, perceptions, and coping strategies. *Journal of Criminal Justice Education*, *10*(2), 231–246. doi:10.1080/10511259900084561

Harley, D. A. (2008). Maids of academe: African American women faculty at predominately White institutions. *Journal of African American Studies*, *12*(1), 19–36. doi:10.100712111-007-9030-5

Hollis, L. (2017). Workplace bullying in the United States and Canada: Organizational accountability required in higher education. *Comparative Civilizations Review*, *76*(76). Retrieved from https://scholarsarchive.byu.edu/ccr/vol76/iss76/13

Jones, T. B., & Osborne-Lampkin, L. (2013). Black female faculty success and early career professional development. Negro Educational Review, 64(1/4), 59-75,135.

Keashly, L., & Neuman, J. H. (2010). Faculty experiences with bullying in higher education. *Administrative Theory & Praxis*, *32*(1), 48–70. doi:10.2753/ATP1084-1806320103

McChesney, J. (2018). *Representation and pay of women of color in the higher education workforce. A CUPA-HR research brief.* College and University Professional Association for Human Resources.

McFarland, J., Hussar, B., Wang, X., Zhang, J., Wang, K., Rathbun, A., … Mann, F. B. (2018). *The Condition of Education 2018.* National Center for Education Statistics 2018-144.

Merritt, D. J. (2008). Bias, the brain, and student evaluations of teaching. *St. John's Law Review*, *82*, 235.

Nast, H. J. (1999). "Sex", "race" and multiculturalism: Critical consumption and the politics of course evaluations. *Journal of Geography in Higher Education*, *23*(1), 102–115. doi:10.1080/03098269985650

O'Meara, K. (2002). Uncovering the values in faculty evaluation of service as scholarship. *The Review of Higher Education*, *26*(1), 57–80. doi:10.1353/rhe.2002.0028

Park, S. M. (1996). Research, teaching, and service: Why shouldn't women's work count? *The Journal of Higher Education*, *67*(1), 46–84. doi:10.2307/2943903

Settles, I. H., Buchanan, N. T., & Dotson, K. (2019). Scrutinized but not recognized: (In)visibility and hypervisibility experiences of faculty of color. *Journal of Vocational Behavior*, *113*, 62–74. doi:10.1016/j.jvb.2018.06.003

Smith, P. J. (1999). Teaching the retrenchment generation: When Sapphire meets Socrates at the intersection of race, gender, and authority. *William & Mary Journal of Women and the Law*, *6*, 53.

Snyder, T. D., de Brey, C., & Dillow, S. A. (2016). *Digest of Education Statistics 2014* (50th ed.). NCES. Retrieved from https://eric.ed.gov/?id=ED565675

Sprague, J., & Massoni, K. (2005). Student evaluations and gendered expectations: What we can't count can hurt us. *Sex Roles*, *53*(11), 779–793. doi:10.100711199-005-8292-4

Steiner, S., Holley, L. C., Gerdes, K., & Campbell, H. E. (2006). Evaluating teaching: Listening to students while acknowledging bias. *Journal of Social Work Education*, *42*(2), 355–376. doi:10.5175/JSWE.2006.200404113

Tierney, W. G., & Rhoads, R. A. (1993). *Enhancing promotion, tenure and beyond: Faculty socialization as a cultural process.* ASHE-ERIC Higher Education Report No. 6. Retrieved from https://eric.ed.gov/?id=ED368322

Tillman, L. C. (2001). Mentoring African American faculty in predominantly White institutions. *Research in Higher Education*, *42*(3), 295–325. doi:10.1023/A:1018822006485

Wallace, S. L., Lewis, A. K., & Allen, M. D. (2019). The state of the literature on student evaluations of teaching and an exploratory analysis of written comments: Who benefits most? *College Teaching*, *67*(1), 1–14. doi:10.1080/87567555.2018.1483317

Ward, K. (2003). *Faculty service roles and the scholarship of engagement*. ASHE-ERIC Higher Education Report. Jossey-Bass Higher and Adult Education Series. Retrieved from https://eric.ed.gov/?id=ED476222

What is DEI? | D5. (n.d.). Retrieved January 3, 2020, from http://www.d5coalition.org/tools/dei/

Yang, Y., & Carroll, D. W. (2018). Gendered microaggressions in Science, Technology, Engineering, and Mathematics. *Leadership and Research in Education*, *4*, 28–45.

Zambrana, R. E., Ray, R., Espino, M. M., Castro, C., Douthirt Cohen, B., & Eliason, J. (2015). "Don't leave us behind": The importance of mentoring for underrepresented minority faculty. *American Educational Research Journal*, *52*(1), 40–72. doi:10.3102/0002831214563063

Chapter 6
A Journey Worth Traveling:
Mentoring and Role Models Matter

Pamela E. Scott-Johnson
California State University, Los Angeles, USA

Pamela M. Leggett-Robinson
PLR Consulting, USA

ABSTRACT

Women of color have historically been underrepresented across the sciences. Neuroscience is no exception. Unfortunately, few studies have examined or shed light on how the dual presence of race and gender affects the educational and professional experiences of African American women in science. This chapter will reflect upon the journey of being an African American woman of science (psychology and neuroscience) in the academy and the blessings not abundantly clear. Through a critical lens, recognizing how the journey would have been more difficult without the supportive network of individual and the critical importance of Historically Black Colleges and Universities. Understanding the context of the times and the need to develop networks that facilitate success of future generations of African American female scholars is crucial.

INTRODUCTION

Education has long been seen as an essential tool of liberation for Black women and girls. Female slaves recognized the importance of reading and writing and thus, remained steadfast in their commitment to the advancement (freedom) of their race and gender through education.

DOI: 10.4018/978-1-7998-4858-5.ch006

In Natchez, Louisiana, there were two schools taught by colored teachers. One of these was a slave woman who had taught at a midnight school for a year. It was opened at eleven or twelve o'clock at night, and closed at two o'clock a.m. Milla Granson, the teacher, learned to read and write from the children of her indulgent master in her old Kentucky home. Her number of scholars was twelve at a time, and when she had taught them to read and write she dismissed them, and again took her apostolic number and brought them up to the extent of her ability, until she had graduated hundreds. A number of them wrote their own passes and started for Canada. (Lerner, 1973, pp. 32-33).

Despite their essential role in the fight for equal access to education, Black women continue to face formidable barriers in achieving post-secondary education, especially in the areas of Science, Technology, Engineering, and Mathematics (STEM). Black women have historically been under-represented across the STEM fields, especially mathematics, computer science, and engineering (House of Commons Science & Technology Committee, 2014). Given that women of color (Black and Brown) will comprise approximately 53% of the U.S. population by the year 2050 (Center for American Progress, 2013), the idea of dismissing or leaving them out of a national conversation on STEM and asking them to blend themselves into other group identities (by gender and/or race) is educationally unsound and inequitable.

As Black girls continue to experience marginalization, oppression, and 'chilly' STEM classrooms, there is an obvious failure on the part of researchers to examine and conceptualize the integrated issues of race and gender (Boston & Baxley, 2007; Mirza, 2009; Pinder, 2008). When researchers examine marginalized groups in STEM education, the focus is almost exclusively on Black males and White females, with little attention devoted to the unique experiences and needs of Black females. (Ricks 2014). Additionally, few studies have examined or shed light on how the dual presence of race and gender affects the educational and professional experiences of Black women in STEM.

... because feminist epistemologies tend to be concerned with the education of White girls and women, and race-based epistemologies tend to be consumed with the educational barriers negatively effecting Black boys, the educational needs of Black girls have fallen through the cracks" (Evans-Winters and Esposito, 2010 p. 12).

Role Models and mentoring are useful in addressing intersectionality of race and gender in the STEM educational process and professional environment for Black girls and women. Both disrupt the narrative many Black females may be internalizing, and provide them access to the wisdom, experiences, and survival skills from their elders. Additionally, both allow Black females to understand and validate their common experiences, and thereby further understand their shared viewpoint. This chapter explores the relationship of intersectionality of race and gender and the importance of role models and mentoring for Black girls (women) in STEM.

EFFECTS OF BROWN *CONTINUE* TODAY

Sixty years after *Brown v. Board of Education of Topeka*, in which the Supreme Court ruled that state laws segregating schools were unconstitutional, K-12 schools are more segregated than they were in 1968. Today, approximately 75 percent of Black students attend racially segregated schools, thus placing Black girls at the bottom of the educational rung (Nelson 2018). Racial and ethnic stigmas are still widely present in *all* educational settings, and they continue to have a strong inñuence on Black girl's perceptions of self, their developing identities and identiðcations, and on their academic performance (Crocker, Major, & Steele 1998; Ogbu 1994; Rowley, Sellers, Chavous, & Smith 1998; Sellers, Chavous, & Cooke 1998; Steele 1997; Steele & Aronson 1995; Steele, Spencer & Aronson 2002; Zirkel 2002, 2004b). Researchers know that these "achievement gaps," or alleged differences in educational performance based on race, are not due to an innate lack of ability in Black female students; but rather, contemporary manifestations of the racial hierarchy (structural arrangements, policies, and practices) that continue in both K-12 and post-secondary institutions (Prime 2018).

An unintended consequence of the fight for integration has been that there are now many fewer Black teachers teaching Black students than there were before 1954. After *Brown*, many Black schools were closed or absorbed by White schools. On account most white communities in the 1950s and 1960s preferred White teachers over Black teachers, court-ordered desegregation ended the teaching careers of Black educators (Toppo & Nichols 2017). The ruling; however, did not account for how their absence would affect the education of students of color. Today, teachers nationwide are overwhelmingly white. According to the most recent data available, white people account for 82 percent of teachers in the United States (Toppo & Nichols 2017; Will 2018). As the proportion of Black, Latino, Asian, Indian, African and other "non-white" students grows inexorably in our nation, the teachers many of these children encounter are nearly all White, thus, the racial K-12 mismatch continues.

Structural Racism and Racial Mismatch in STEM K-12 and Post Secondary Education

Racial mismatch has a profound impact on Black girls and women in STEM K-12, STEM higher education, and the STEM workplace. The dual presence of race and gender affect the educational experiences of Black girls and women because so often the experiences are addressed in isolation. The manifestations of racial mismatch are the reluctance, exclusion, mis-opportunities, and "under-education" of Black female students within the K-12 and higher education system (Catsambis 1994; Riegle-Crumb et al 2010). In an effort to understand the impact of racial mismatch in education, especially as it pertains to STEM education, an examination of STEM K-12 and STEM higher education social structure is paramount because the social, psychological, and structural dimensions influence student identity, belonging, self-efficacy, and encouragement (Leggett-Robinson et al 2018). Both institutions (K-12 and Higher Education) operate such that they are aligned similarly - the way they distribute and convey information about STEM culture, assign meaning to STEM academic success, and produce and/or reproduce racial disparities.

Structural racism is a social system of hierarchy whereby economic, educational, political, and social ideologies routinely advantage White people at the expense of Black, Latino, Asian, Pacific Islander, Native American, Arab and other racially oppressed people (Bonilla-Silva 1997; Lawrence & Keleher 2004). It is the discriminatory treatment, unfair policies, and inequitable opportunities and impacts, produced and perpetuated by institutions (schools, mass media, etc.). It is distinguished from the bigotry or racial bias of individuals by the existence of systematic policies and practices within institutions (Leggett-Robinson & Villa 2019). Individuals within these institutions take on the power of the institution when they act in ways (consciously or unconsciously) that advantage and disadvantage people, based on race. Structural racism encompasses entire educational systems and is the most profound and pervasive form of racism.

STEM can be defined as a structural racism of sorts, as it remains a common occurence to everyday, ordinary interactions in our educational system. Structural racism within STEM education encompasses: (1) history, which lies underneath the surface, providing the foundation for white academic advantage, (2) culture, providing the normalization and replication of racism and, (3) interconnected institutions, school systems, and policies, the key relationships and rules across an intellectual society which provide the legitimacy and reinforcements to maintain and perpetuate racism. The clear indicators of STEM structural racism are inequalities in power, access, opportunities, treatment, and policy impacts and outcomes (Lawrence & Keleher 2004). Unfortunately, STEM structural racism is more difficult to pinpoint in a particular K-12 school or higher education institution because it involves the reinforcing effects of multiple schools, institutions, and cultural norms, past and present. From structural racism emerges institutional, interpersonal and internalized racism.

In STEM education, the system of structural racism takes the form of educational opportunities, practices, and beliefs based on race and gender and cultural values around academic expectations of math & science preparedness. The culture of STEM education has traditionally been characterized by a "survival of the academically fit" and attributes any failure in STEM to student characteristics (Gasman 2009). A survival-of-the-fittest mindset implies Black girls (women) are responsible for their own under-representation, and it sets the stage for stereotypic assumptions that many Black female STEM students report experiencing in K-12 and post-secondary education (Brown et al 2016; McGee 2016). This mindset contributes to presumptions of Black female students' academic inferiority and lack of belonging in STEM.

STEM behaviors, practices, and beliefs are readily apparent in K-12 school communities and STEM departments in higher education. These perspectives often deteriorate an educator's expectations for Black female students and weaken the educator's capacity to recognize gifts, talents, and extraordinary abilities in diverse forms (Ford & Grantham 2003; Osanloo et.al 2016). This belief system persists mostly in the conscious thought (of teachers and professors) but surfaces when differences between those deemed as "other" are seen as academic deficits in the STEM culture. The belief system acting in tandem with the racial mismatch of STEM educators perpetuates oppressive, dominating behaviors and practices that undermine the educational experience of many Black female students who chose to pursue STEM pathways. Additionally, these racial mismatch interactions are experienced on a one-to-one teacher-student level as well as operationalized as larger cultural constructs within the STEM education systems of K-12 and higher education and result in the absence or low numbers of Black women in STEM.

Low numbers of Black women in STEM affect more than the women themselves; the numbers represent missed opportunities for STEM fields (Milgram, 2011). Women bring a unique perspective that shapes and influences STEM disciplines and benefits society. On account more than 28% of the nation's population are persons of color, the absence of this group in STEM stifles innovation. When women of color (for this chapter, Black women) are not involved in the design of STEM products, services, and solutions, the needs and desires unique and germane to all women may be overlooked and the resulting solutions not designed to represent all users (2016 Census Bureau, Hill et al., 2010).

IMPORTANCE OF ROLE MODELS

The lack of Black female STEM K-12 teachers and STEM professors removes a source of visible role models for existing and aspiring female scientists. According to Dennehy and Dasgupta (2017), having academic contact with Black female professionals in STEM has positive influences on Black female students. For Black girls and young women in STEM, being able to identify and engage with role models who share their racial identity helps to not only imagine themselves as STEM professionals but also enhance their perceptions of STEM careers, boost self-efficacy, increase motivation, and promote STEM belonging. At every stage of the STEM journey for Black girls and young women, these role models provide vital support. Although few studies have reported Black girls and women can excel in STEM *without* role models, many will encounter problems envisioning certain positions as viable possibilities, even if they have some intrinsic interest in the subject matter.

Stereotypical images of scientists are perpetuated when scientists are constantly depicted as a white man wearing a lab coat with bifocals arched over the brim of his nose as he constructs experiments in his laboratory (Ladson-Billings, 1997; Van Langen & Dekkers, 2005; Powell, 1990; Riegle-Crumb et al., 2010). As with any stereotype, these images are "cognitive structures that contain the perceiver's knowledge, beliefs,

and expectations about human groups" (Peffley et al 1997). Thus, pictures of Aristotle, Albert Einstein, Sir Isaac Newton, Benjamin Franklin, Bill Nye the Science Guy, Dr. Jekyll and Mr. Hyde, and Victor Frankenstein solidify the *accepted* image of scientists and further demonstrate that whether in fiction or in life, scientists are often depicted as White and male (Farinde & Lewis, 2012).

Dominant images, not resembling Black female students, engenders a false assumption that these young girls should not aspire to become scientists, because either STEM careers are only designated for White men, or the obtainment of a STEM career would go against accepted traditional gender role assignments for Black women. These ideas are subtly implied through the absence of images of Black female scientists both within society and the American education system as well as the absence of STEM K-12 teachers and higher education professors. If Black girls cannot visualize themselves in STEM careers because they have never seen women in those positions, they will be much less likely to ever use their innate aptitudes and abilities in a math- or science-oriented profession. Instead, the STEM education system will continue to view them as intellectually and academically hindered; consequently, the culture, bias, and prejudice ideologies will continue. According to Milgram (2011), "Women and girls need to see female role models in the workplace that look like them—over and over and over again" (p. 5). This is especially true for Black females as their dual presence of race and gender affects every aspect of their STEM journey; influencing experiences and opportunities.

Fixed Mindset vs. Growth-Mindset

Black female role models in STEM disrupt systems of implicit racial and gender bias by providing a safe space to promote positive ethnic and racial identity within the STEM culture. They also help Black female students develop a growth-mindset versus a fixed-mindset (Hill et al 2010; TPT 2013).

In a fixed mindset, people believe basic qualities, like intelligence or talent, are simply fixed traits. They also believe that talent alone creates success, without effort. In a growth mindset, people believe basic abilities can be developed through dedication and hard work; brains and talent are just the starting point. Bias and prejudice ideologies within the STEM culture encourage a "fixed" mindset; that is, the belief - intelligence is simply something you are either born with or without. The students who show promise earlier on in STEM without having to work as hard are encouraged more from the beginning and throughout their STEM educational journey and professional pathway. This behavior leads to encouraging only these "types" of students, thus leaving students who don't fit the "fixed mindset" stereotype to feel as though they don't belong in STEM. Additionally, students of color with a fixed mindset are:

... more likely to believe stereotypes, lose confidence, and disengage from STEM as a potential career when they encounter difficulties in their coursework" (Hill et al 2010, p. 21).

Despite the pervasiveness of the idea that STEM skills are innate, discoveries in science are often a product of hard work and collaboration. Decades of research have shown that a growth mindset in STEM leads to more effective teachers, learners, school systems, as well as creates a culture of inclusion and diversity in our STEM organizations and institutions (Sundermier 2018). These two mindsets lead to different learning processes and outcomes, causing Black females to engage in STEM learning environments very differently (Sundermier 2018; Dweck 2008). When a STEM field or a particular STEM department, research center or firm espouses a fixed mindset, research shows that female marginalized groups feel less trust and less belonging in that organization (Dweck 2008). Conversely, when a growth mindset is elevated over a fixed mindset, trust and belonging are foundational and female marginalized groups flourish. They feel respected and valued for their contributions and feel like they have a chance to learn and grow.

Black female role models in STEM help to develop and model a growth mindset for ***all*** students. Research by Hill et al. (2010) found that fostering a growth mindset helps shield young women against negative stereotypes and bias. Additionally, the development of a growth mindset model, increases STEM self-efficacy.

When a girl believes that she can become smarter and learn what she needs to know in STEM subjects—as opposed to believing that a person is either born with science and math ability or not—she is more likely to succeed in a STEM field" (Hill et al, 2010, p. 25).

STEM Belonging

Sense of belonging refers to a students' sense of being accepted, valued, included, and encouraged by others (teachers and peers) in the academic classroom setting (Rainey et al 2018; Goodenow 1993). Literature indicates that a sense of belonging in STEM has a significant impact on educational success and persistence, especially for Black female students. A sense of belonging in STEM is associated with academic achievement, retention, and persistence in college and these impacts are frequently more pronounced for students from marginalized groups (Strayhorn 2012). Black female STEM students have consistently reported less sense of belonging than their male and White female counterparts, felt more pressure to conform to masculine norms, and worked harder to establish their claims of scientific authority to participate in STEM (Good et al 2012; Johnson 2012; Smith et al 2013; Hirshfield 2011). The perception of having to exert additional effort than their male and/or White female counterparts to belong and succeed oftentimes decreases motivation to pursue a STEM field (Smith et al 2013). The degree of fit or perceived belonging between an individual's social contexts and her personal characteristics influences not only the nature of social interactions but also the makeup of attitudes and beliefs about self and others. Feelings of "belonging uncertainty" and fear of confirming negative stereotypes compromise the academic and/or professional performance of Black women in STEM.

Although an important impact of the structured racism within the STEM culture, there are few studies that look at the experiences of Black girls and women regarding belonging (Johnson 2012; Mickelson et al 2015). Studies about belonging in STEM fields that discuss racial identity often omit a gender analysis, thereby excluding Black females from the literature. Lastly, intersectional works that investigate the Black female experience (e.g., Carlone and Johnson 2007; Crenshaw 1991), don't focus on their sense of belonging in STEM.

The dearth of role models of similar gender and race further undermines Black girls' and women's senses of belonging in STEM. It is through exposure to role models who share gender and racial identities that young women entering STEM fields may be protected to some extent from the barriers and challenges and begin to see their belonging in the STEM field (Leggett-Robinson & Villa 2019; Dasgupta 2015).

IMPORTANCE OF COMMUNITY (HBCUS, MENTORS, AND PEERS)

Race outweighs or trumps other identities, because through race, American society initially identifies and defines people (Singleton 2015). The same is true in the STEM culture. Although women have difficulties navigating the STEM landscape, White women do not possess a racial identity presupposed to intellectual and academic inferiority. Thus, Black women's struggles within STEM are often different and the need for mentoring and community (with same racial and gender identity) is important to their STEM academic and professional success. Regarding the STEM culture, White is the dominant race and male is the dominant gender; moreover, it represents the "standard" by which STEM experiences and perspectives of "others" are compared and judged.

Communities

The link between identity, culture, and learning is a fundamental endeavor for researchers conceptualizing how STEM identity formation, race, and gender intersect in academic and professional development within the STEM culture. STEM persistence and retention (academia and workplace) are dependent not only on motivation and interest but are also associated with the ability to identify with STEM and STEM careers (Carlone & Johnson 2007).

Advances in STEM career development theory suggests that the most successful intervention strategies for women promote STEM identity (Graham, 2013; Syed, 2011; Estrada, 2011). STEM identity is a reflection of how one understands and positions oneself within the STEM culture. It can be defined as the composition of self-views that emerge from socialization into the norms and practices of STEM and is ultimately built from the internalization of positive or negative experiences. For Black women in STEM, who may simultaneously find themselves at the intersection of race and gender, STEM identity experiences are particularly complex (Carlone & Johnson 2007). Developing a strong STEM identity involves Black female students coming to see themselves as a scientist and feeling a sense of belonging in the scientific community (Thiry 2011; Hernandez 2013; Perez 2014). According to Arnett (2000), particular types of adult support are necessary to promote students' development of resilience, facilitate learning to adaptively cope in different domains, and develop STEM identity.

Settings themselves have the power to signal the degree of threat or safety a person will experience. For Black women, STEM settings create the expectation that a person's treatment will be contingent on one of their social identities. Social identity contingencies are possible judgments, stereotypes, opportunities, restrictions, and treatments that are tied to one's social identity in a given setting (Steele, Spencer, & Aronson 2002). Purdie-Vaughns (2008) asserts situational cues trigger a person's expectations about the kinds of social identity contingencies they may face, and these expectations affect whether they can trust and feel comfortable in a given setting. For instance, when Black women expect their social identity contingencies in STEM to be positive or neutral, the setting (community) can be characterized as identity-safe. Because Black women's STEM identity values are tied to experiences within these specific settings, this group draws information from situational cues that hold relevance for the value and the status associated with their group.

Positive communities are deemed identity-safe and create culturally patterned structural supports (knowingly and unknowingly) to help Black women succeed in STEM. In STEM communities at Spelman or Bennett, where the dominant race and gender is Black and female, STEM experiences and perspectives of Black women are embraced and nurtured. These positive communities capture interactions and synergistic influences of social support (i.e., mentoring) to promote STEM identity, motivation, and persistence.

Mentoring

Mentoring is a relationship in which a more experienced individual guides and advises a less experienced individual by providing instrumental and psychosocial support. Both informal and formal mentoring relationships have the potential to alleviate stress by increasing the mentee's self-confidence, forewarning of academic or career stress, and suggesting ways to deal with adversity. In addition, it is suggested that female mentors provide unique role models for female students or professionals because they can relate to the causes of stress unique to females. Particularly in STEM, mentoring has been repeatedly cited as an effective method for recruiting and retaining women of color in STEM fields (Settles et al 2007; Hewlett et al 2008; Wilson et. al 2010; Yeager 2000; Terenzini & Wright 1987; Van Eps et al 2006; Abriam-Yago 2002; Tinto 1993; DeFour & Hirsch 1990). Black women serving as mentors for female students provide critical socialization experiences thus affecting persistence, post-graduation, and workplace success. A lack of mentoring for Black women in STEM results in less access to professional networks, decrease in STEM identity and belonging, and a lack of diversification in STEM fields.

Mentoring can be a powerful lifelong learning process, facilitating the intergenerational transfer of knowledge and "know-how" (Murrell et al 2008). Mentoring relationships are beneficial and essential to the academic and professional development of Black female scientists and engineers, but they do not naturally occur for all. Where geographic, cultural, academic, economic, and other differences create barriers to identification of those commonalities, i.e., STEM culture, mentoring relationships are less likely to form.

Black women mentors often offer emotional support through friendship and Black female students have reported that they have a higher level of trust with a Black woman mentor. Black women mentors are believed to be more relatable, trustworthy, and understanding than White mentors, male or female. Black women often feel that they cannot truly express themselves with White mentors or show any weaknesses, whether academic or personal. More importantly, Black female STEM professionals serving as mentors reported increases in their own self confidence as a result of their mentoring experiences, and Black female mentors also highlighted perceptions of improvement in their skills for recruiting new talent to their fields as a result (Muller et al 2012). These examples reiterate the importance and significance of the mentoring relationship within the Black STEM community.

HBCU Community

Historically Black Colleges & Universities (HBCUs) play a critical role within US higher education (Mack, Rankins, and Woodson 2013; Mack et al. 2010; Mack, Rankins, and Winston 2011), especially within the STEM disciplines. Archival evidence shows HBCUs to provide more inclusive and nurturing environments for students compared to predominantly White institutions (PWIs) as well as provide a strong sense of belonging through a family and community atmosphere, and similar cultural experiences (Harper & Antonio 2008; Fleming 1984; Allen 1992). HBCU environments facilitate the development of a student's self-concept, self-efficacy, and self-esteem, thereby validating a student's presence and purpose at the institution, and life in general (Davis 1994). Students attending HBCUs are welcomed by a nurturing environment that provides critical resources to overcome academic, social, and financial hurdles.

HBCUs represent 3 percent of the nation's population of higher education institutions, despite relatively small endowments and low institutional resources. However, HBCUs have remained among the nation's top educators of Blacks who go on to earn STEM doctoral degrees and an important producer of Black faculty in STEM fields (Solorzano, 1995; Strayhorn, 2008; Strayhorn, Williams, Tillman-Kelly, & Suddeth, 2012; Perna 2001). Moreover, research attests undergraduates attending HBCUs were more likely to enroll in a graduate STEM program than those attending a PWI (Wenglinsky 1997).

Substantial barriers (discussed throughout the chapter) limit the attainment of Black women in STEM fields even before enrolling in higher education. Despite the importance of K-12 STEM preparation, experiences of fixed-growth mindset, stereotype threat, and feeling of non-belonging, research suggests that institutional structures, policies, and practices at PWIs contribute equally to the low attainment of women and minorities in STEM fields. In particular, colleges and universities that serve predominantly Black populations and/or women appear to be disproportionately effective in promoting the educational attainment of these groups overall, and in STEM fields in particular (Kim and Conrad 2006). The National Science Foundation (2006) reported HBCUs produced 46 percent of black women who earned degrees in STEM disciplines between 1995 and 2004. Research by Thompson (2008) found evidence that many female graduate students and faculty at majority institutions gave credit to their HBCU experience for their STEM preparedness. Conversely, while the campus culture at HBCUs provides Black female students in STEM with a sense of community, self-confidence, and self-efficacy, majority institution STEM graduate programs and the STEM workforce provide a stark contrast. Relative to their share of degree attainment, Black females are underrepresented in the overall context of the STEM workforce (industry, government, and academia).

Additional research suggests the historical and current role of women's HBCUs in promoting the attainment of women in science and engineering careers. HBCUs for women, mitigate the negative impact of the aforementioned barriers by providing a cooperative rather than competitive peer culture, promotion of STEM identity, belonging, and self-efficacy, and community of mentors and role models. Compared with their STEM counterparts, who attend PWIs, Black female students who attend HBCUs experience less social isolation, alienation, personal dissatisfaction, and overt racism (Harper et al. 2004; Pascarella and Terenzini 2005). HBCUs for women provide a social, cultural, and racial environment that is more supportive, caring, and nurturing for Black females. Moreover, HBCUs for women embody a STEM community that acknowledges and celebrates the intersection of race, gender, and STEM as well as reflects the assumption that all women can succeed in STEM fields.

STEM REFLECTION

According to Buhrman (2006), there are three major points of influence on Black women's STEM career decisions: parents, peers, and educators. In this reflective narrative, I think about the influence of family, peers, and Spelman College on my journey in STEM. At each phase of the journey, I was blessed with role models, mentors, and community that helped to develop a STEM identity, feelings of STEM belonging, and a belief I could be successful in STEM as a Black woman.

I grew up in Newberry, SC with my mother and father and two siblings (Willie, Jr. and Nelson). I am the only girl, and in the middle. Both of my parents (the late Willie L. Scott, Sr. and the late Gloria Nelson Scott) were high school teachers, with my mother having taught chemistry, math, and physics, and my father psychology. Both of my parents were coaches. My father coached football, basketball, and track; mom coached the girls' basketball team and cheerleading. I think it important to note that when my parents first entered the teaching profession, they did so in a separated school system in the south. Both began careers in the segregated school system in Newberry, SC. When the schools were integrated, my mother taught only mathematics. In fact, she was "asked" to teach general mathematics, rather than algebra, calculus, advanced mathematics and other STEM courses that she had taught during her first 20 years in the system. These college-bound STEM courses were reserved for the White teachers. Mom recognized that she could make sure that her Black students (and poor White students) would have someone who would look out for them, and who would care about their progress. Since my parents lived and worked in Newberry, they were familiar with the families of many of these students and provided context and support for the community during and after segregation. In addition to my mother's teaching math at the high-school level, my grandmother's sister, Mrs. Ruby Sligh, taught sixth grade math (algebra) and was the lead math teacher for the re-named Gallman Middle School. Given the talent in my family and the commitment to education, I did not have a fear of mathematics or science.

The Spelman Experience

I remember having a conversation with my father about college and wanting to attend Spelman. Although my father's sister-in-law Mrs. Mary Reeder Abrams (aunt) was a staunch advocate of supporting my attending Spelman, it was my music teacher (Mrs. James Myers) who invited me to attend the Annual Spelman-Morehouse Christmas Concert in support of her former student and family friend (Dr. Ave Maria Lindsay Marshall). We met her before the concert, walked the campus, and then went to the concert. I was hooked, so much so that Spelman was the only college I applied to.

My early college ambition was to become a lawyer, then the first Black female governor of South Carolina (*Let's say that I have not made it to the governor's mansion*). Sometime during my four-year college matriculation, I changed my major from political science to psychology; however, my academic advisor encouraged taking math courses for electives. My elective courses included biostatistics, math courses for majors, and graduate statistics. I spent as much time with the math faculty as I did the psychology faculty, such that the math faculty thought I was a math major. I also was encouraged and invited to apply to the newly created Minority Access to Research Careers (MARC) program, funded by the National Institutes of General Medical Sciences with Dr. Geraldine Woods (Head of the Institute) and Elwood Bynum (Program Director). I discovered my love for research through the MARC program while at the same time, I was discovering neuroscience and physiological psychology. I realized that I could be a psychologist without having to be a clinician. The MARC program, along with my broader leadership and academic training gave me confidence and strength. I soon realized I was part of a cohort of change makers; I was educated, a woman, and African American. I add that my university president, the late President Donald M. Stewart, was also one of my supporters, and encouraged me to pursue every opportunity Spelman provided.

Upon graduation, I left the supportive environment of Spelman College to attend Princeton University, an PWI that had only begun to enroll women just a few years earlier than my entry as a graduate student.

The Graduate School Experience

I will acknowledge that graduate school was difficult, not only because my advisor believed that his liberalism was a favor to the world but also because it took me a while to adjust to an environment with people who continued to question my intellect. I remember wanting to submit an abstract to the regional conference. My advisor supported and encouraged the undergraduates in the lab and thought that I needed more research. This same advisor once told me to think and perform like one of his "elite" undergraduate students. He said this to me on more than one occasion. I can only tell you that I was emotionally devastated each time he uttered the words. In fact, I would hate having meetings with him because the meetings left me emotionally drained and feeling completely incompetent. I started scheduling them on Fridays so that I would have the weekend, which included attending Sunday School and services, to recover.

As I talked to other Black graduate students in the department (because there were no other Black graduate students in neuroscience) none of them seemed to be having the intensity of my experience. There were three Black females who were already in the department. In fact, one of my colleagues shared my experience with her mentor, and he became my guide. Two of them were older, had been in the workforce prior to graduate school, and had families. Their perspectives were much different due to their familial priorities. The other Black female indicated that she was better off than me because she did not graduate from a historically black college. Of the four of us, only two graduated. Despite my difficulties; however, I was one of the two. The student that touted her "position" of being better due to pedigree, did not.

Three incidents helped to forge my pathway to graduation. The first was when I got mad enough at my situation that I reminded my advisor that I was a graduate student and not an undergraduate, and that making statements that belittle my intelligence were insulting. The second incident was when a faculty member wrote a recommendation for a MARC pre-doctoral fellowship that was more of a commentary on affirmative action than on my abilities or the work I performed. The agency leadership was so concerned about the letter that they called me and asked if I was ok. At that time, I was not aware that I was allowed to receive a copy of my letter prior to it being submitted or that I was actually in a hostile environment. Both incidents forced me to remove myself from my original lab and seek out a new group. Through the help of the graduate director and two other faculty members, the transition to the new group was smooth and I was able to recover emotionally. In my new group, I learned how to navigate and manage my interactions with other faculty members in my area as well as how to engage and not isolate. I also learned how to be a scientist and stand my ground for my ideas. My second advisor, my colleague's advisor, and the graduate director, gave me the support and mentored me the same as others in their lab. I was able to share my thoughts and garnered support from the department. The third incident occurred during the annual neuroscience and psychology meetings. At these meetings, I met other African Americans (attending both HBCUs and PWIs). I quickly began to realize that nothing was wrong with me, and that there was support and conversation about science that could be beneficial to my project. From that point, I began to attend the women in neuroscience gatherings. I realized the power of being in the same space as others that "looked like me" and that could provide a safe space to discuss professional and personal accounts. These three incidents showed me that role models and mentoring could occupy informal and formal spaces with people that looked like me or not. It also opened my eyes to ***my resilience***.

Realization of Being a Black Woman in the (STEM) Professoriate

I realize that the nurturing I received from my family and from my undergraduate institution prepared me for the journey that required serious resilience. Over the years, I tried to provide the same nurturing for those who I mentored and trained. You can't imagine my disappointment when I met one of my protégés who shared similar experiences I had had in my lab. She described an incident whereby her advisor accused her of not having the intellectual capacity to write a draft of a book chapter. Given the efforts to increase representation in the field of neuroscience, I was shocked and surprised. Through this exchange I realized, every black woman needs and deserves to have *an advocate, a listener, and a motivator*, and a conversation that reminds us of and reinforces our value.

After having a career in industry for 10 years, I began my academic career at my alma mater. Again, I had the benefit of having women nurture me and was fortunate to have mentors. Mentorship was more important than I realized as I encountered STEM departments with faculty demographics not reflective of the students and disciplinary bullying. The more bench-top STEM productivity, the more entitled faculty felt. For example, I was trained in neuroscience. My training in psychology and neuroscience seemed to be less STEM than those trained in biology. I soon realized there was little respect for the research that occurred further away from equipment and bench-related experimentation. The social sciences were becoming more and more feminized in their participation, thus the culture and faculty demographics of STEM at my alma mater were beginning to look like that of a PWI.

Great Mentors

I mentioned that I had the privilege of having mentors who encouraged me to stay on track. I entered Spelman College in 1978. It was not until my sophomore year that I declared psychology as my major. When I entered I thought I would major in political science. My goal then was to be the first Black female governor of South Carolina. Although I had exceptional mentoring in the political science department, it was my classmate and sorority sister who told me about the MARC program. My psychology advisor, Dr. Oran Wendall Eagleson, encouraged and supported my application. The program was a new initiative to encourage Black students to consider careers in biomedical sciences. It took me a while to find out what biomedical science was, but after attending a number of information sessions, I realized that I had an opportunity to do something really different, and that psychology was my chosen field. MARC was a newly established program by the National Institute of General Medical Sciences (NIGMS). Dr. Geraldine Woods, the first African American woman appointed to the National Advisory Council for the NIGMS, worked tirelessly to establish the program. In 1980, I was in the second cohort of students funded through the program. Armed with supportive and knowledgeable advisors like Dr. Eagleson, Dr. Etta Falconer, and Dr. Nagambal Shah, I took the necessary courses to prepare me for the biomedical program. Additional mentors and role models included my MARC mentor, Dr. Dwayne Jackson, Dr. Margaret Weber-Levine and her late husband Steven Levine as well as the director of the MARC program for the Atlanta University Campuses, Dr. Isabella Finkelstein. My participation in MARC led me to Washington DC where I met the two people responsible for MARC, Drs. Geraldine P. Woods and Elwood Bynum. My mentors helped to secure an internship during my senior year at National Institute of Health (NIH) with Dr. Wallace Mendelson and the late Dr. Christine Gillin. I returned to Spelman feeling confident and prepared for graduate training. Because I had been exposed to a new research area, I applied to graduate programs in neuroscience in the area of sleep or somnology without fear. I had garnered enough confidence, with the support of my mentors and university president, President Donald M. Stewart, to apply to Cornell University, Princeton University, and the University of Chicago.

FINAL THOUGHTS OF MY STEM JOURNEY

There is a saying that goes, "There but for the grace of God, go I." Here's what I see. Had I not had the parents I had, with a mother who taught chemistry, physics, and mathematics; had I not had a great aunt who taught mathematics in middle school; had I not attended Spelman College, an institution with an expressed purpose of developing African American women as leaders; had I not had great mentors who encouraged my engagement in research and STEM; had Geraldine Woods and Elwood Bynum not fought for, established, and directed the MARC program, I would not have had the tenacity, confidence, and support and would not have a STEM career, nor would I feel obligated to make sure that others have better mentoring and STEM opportunities.

The nurturing environments of Newberry, SC and Spelman College were critical for a small-town girl to become a powerful woman leader in STEM. Each day I hold onto the tools of STEM belonging, STEM identity, growth mindset, and mentorship so graciously given to me as a young girl, and as a young college-aged woman.

The journey in STEM and in academia is hard because no one wants to talk about the ways the academy attempts to homogenize itself by keeping unrealistic, culturally irrelevant criteria (e.g., GRE scores, publication expectations without regard to institutional type and resources), and ideologies consistent with the founding fathers of the academy (e.g. tattered relationships, intellectual/disciplinary bullying, or expectations for work hours and burnout). In the era of difficult conversations and diversity/equity/inclusion, now is the time to start anew and have integrity to move forward in a direction that enables and empowers marginalized groups in STEM.

Regardless, we will not let barriers deter us in providing support and mentorship for the next generation. We will speak up, mentor, train, and care. It's the gift we have because of the gift we have been given. We matter in bridging the next generation. Mentoring allows us to maintain our connections as Black women, mentors, role models, and STEM leaders.

REFERENCES

Abriam-Yago (2002). Mentoring to empower. *Minority Nurse*, 64-65.

Allen, W. R. (1992). The color of success: African-American college student outcomes at predominantly white and historically black public colleges and universities. *Harvard Educational Review*, *62*(1), 26–44. doi:10.17763/haer.62.1.wv5627665007v701

Arnett, J. (2000). Emerging adulthood: A theory of development from the late teens through the twenties. *The American Psychologist*, *55*(5), 469–480. doi:10.1037/0003-066X.55.5.469 PMID:10842426

Bonilla-Silva, E. (1997). Rethinking racism: Toward a structural Interpretation. *American Sociological Review*, *62*(3), 465–480. doi:10.2307/2657316

Boston, G. H., & Baxley, T. (2007). Living the Literature Race, Gender Construction, and Black Female Adolescents. *Urban Education*, *42*(6), 560–581. doi:10.1177/0042085907305186

Brown, B. A., Henderson, J. B., Gray, S., Donovan, B., Sullivan, S., Patterson, A., & Waggstaff, W. (2016). From description to explanation: An empirical exploration of the African American pipeline problem in STEM. *Journal of Research in Science Teaching*, *53*(1), 146–177. doi:10.1002/tea.21249

Buhrman, J. (2006). Extraordinary Women Engineers Project (EWEP). *Mathematics Teacher*, *100*(4), 283.

Carlone, H. B., & Johnson, A. (2007). Understanding the science experiences of successful women of color: Science identity as an analytic lens. *Journal of Research in Science Teaching*, *44*(8), 1187–1218. doi:10.1002/tea.20237

Catsambis, S. (1994). The path to math: Gender and racial-ethnic differences in mathematics participation from middle school to high school. *Sociology of Education*, *67*(3), 199–215. doi:10.2307/2112791

Census Bureau. (2016). https://factfinder.census.gov/faces/tableservices/jsf/pages/productview.xhtml?src=bkmk

Center for American Progress. (2013). Retrieved from https://www.americanprogress.org/

Crenshaw, K. (1991). Mapping the margins: Intersectionality, identity politics, and Violence against women of color. *Stanford Law Review*, *43*(6), 1241–1299. doi:10.2307/1229039

Crocker, J., Major, B., & Steele, C. M. (1998). Social stigma. In D. T. Gilbert, S. T. Fiske, & G. Lindzey (Eds.), *The handbook of social psychology* (4th ed., pp. 554–594). Boston: McGraw-Hill.

Dasgupta, N., Scircle, M. M., & Hunsinger, M. (2015). Female peers in small work groups enhance women's motivation, verbal participation, and career aspirations in engineering. *Proceedings of the National Academy of Sciences of the United States of America*, *112*(16), 4988–4993. doi:10.1073/pnas.1422822112 PMID:25848061

Davis, J. (1994). College in black and white: Campus environment and academic achievement of African American males. *The Journal of Negro Education, 63*(4), 620–633. doi:10.2307/2967299

DeFour, D., & Hirsch, B. J. (1990). The adaption of Black graduate students: A social Network approach. *American Journal of Community Psychology, 18*(3), 487–503. doi:10.1007/BF00938119 PMID:2264561

Dennehy, T. C., & Dasgupta, N. (2017). Female peer mentors early in college increase women's positive academic experiences and retention in engineering. *Proceedings of the National Academy of Sciences of the United States of America, 2017*(May), 5964–5969. doi:10.1073/pnas.1613117114 PMID:28533360

Dweck, C. S. (2008). *Mindset: The New Psychology of Success*. New York: Ballantine Books.

Estrada, M., Woodcock, A., Hernandez, P. R., & Schultz, P. W. (2011). Toward a model of social influence that explains minority student integration into the scientific community. *Journal of Educational Psychology, 103*(1), 206–222. doi:10.1037/a0020743 PMID:21552374

Evans-Winters, V. E., & Esposito, J. (2010). Other people's daughters: Critical race feminism and black girls' education. *Educational Foundations, 24*(1), 11–24.

Farinde, A., & Lewis, C. (2012). The Underrepresentation of African American Female Students in STEM Fields: Implications for Classroom Teachers. *US-China Education Review B 4*, 421-430. https://files.eric.ed.gov/fulltext/ED533550.pdf

Fleming, J. (1984). *Blacks in college: A comparative study of student success in black and white institutions*. San Francisco, CA: Josey-Bass.

Ford, D. Y., & Grantham, T. C. (2003). Providing access for culturally diverse gifted Students: From deficit to dynamic thinking. *Theory into Practice, 42*(3), 217–225. doi:10.120715430421tip4203_8

Gasman, M., Perna, L. W., Yoon, S., Drezner, N. D., Lundy-Wagner, V., Bose, E., & Gary, S. (2009). The path to graduate school in science and engineering for underrepresented students of color. In L. Santiague (Ed.), *Standing on the outside looking* (pp. 63–81). Sterling, VA: Stylus Publishing.

Good, C., Rattan, A., & Dweck, C. S. (2012). Why do women opt out? Sense of belonging and women's representation in mathematics. *Journal of Personality and Social Psychology, 102*(4), 700–717. doi:10.1037/a0026659 PMID:22288527

Goodenow, C. (1993). Classroom belonging among early adolescent students: Relationships to motivation and achievement. *The Journal of Early Adolescence*, *13*(1), 21–43. doi:10.1177/0272431693013001002

Graham, M. J., Frederick, J., Byars-Winston, A., Hunter, A. B., & Handelsman, J. (2013). Increasing persistence of college students in STEM. *Science*, *341*(6153), 1455–1456. doi:10.1126cience.1240487 PMID:24072909

Harper, S., & Antonio, A. (2008). Not by accident: Intentionality in diversity, learning, and engagement. Creating inclusive campus environments for cross-cultural learning and Student engagement. Washington, DC: National Association of Student Personnel administrators.

Harper, S. R., Carini, R., Bridges, B., & Hayek, J. (2004). Gender differences in student engagement among African American undergraduates at Historically Black Colleges and Universities. *Journal of College Student Development*, *45*(3), 271–284. doi:10.1353/csd.2004.0035

Hernandez, P. R., Schultz, P. W., Estrada, M., Woodcock, A., & Chance, R. C. (2013). Sustaining optimal motivation: A longitudinal analysis of interventions to broaden participation of underrepresented students in STEM. *Journal of Educational Psychology*, *105*(1), 89–107. doi:10.1037/a0029691 PMID:24273342

Hewlett, S. A., Luce, C. B., Servon, L., Sherbin, L., Shiller, P., Sosnovich, E., & Sumberg, K. (2008). The Athena Factor: Reversing the brain drain in science, engineering and technology. *Harvard Bus. Rev.*

Hill, C., Corbett, C., & St Rose, A. (2010). *Why so few? Women in science, technology, engineering, and mathematics*. Washington, DC: American Association of University Women.

Hirshfield, L. E. (2011). *Authority, expertise, and impression management: Gendered professionalization of chemists in the academy* (Dissertation). University of Michigan. https://deepblue.lib.umich.edu/bitstream/2027.42/89796/1/lhirshf_1.pdf

Johnson, D. R. (2012). Campus racial climate perceptions and overall sense of belonging among racially diverse women in STEM majors. *Journal of College Student Development*, *53*(2), 336–346. doi:10.1353/csd.2012.0028

Kim, M., & Conrad, C. (2006). The impact of historical Black colleges and universities on the academic success of African-American students. *Research in Higher Education*, *47*(4), 399–427. doi:10.100711162-005-9001-4

Ladson-Billings, G. (1997). It doesn't add up: African American students' mathematics achievement. *Journal for Research: Crossing Boundaries in Search of Understanding*, *28*(6), 697–708. doi:10.2307/749638

Lawrence, K., & Keleher, T. (2004). Structural racism. *Race and Public Policy Conference Proceedings*.

Leggett-Robinson, P., Davis, N., & Villa, B. (2018). Cultivating STEM identity and belonging through civic engagement: Increasing student success (self-efficacy and persistence) for the two-year college STEM student. *Science Education and Civic Engagement*, *10*(1), 24–34.

Leggett-Robinson, P., & Villa, B. (2019). Navigating the landscape of the STEM professoriate: Reflections and insights from women of color. In *Women's Influence on Inclusion, Equity, and Diversity in STEM Fields*. IGI Global Publishing. https://www.igi-global.com/book/women-influence-inclusion-equity-diversity/218525

Lerner, G. (1973). *Black women in white America: A documentary history*. New York, NY: Vintage.

Mack, K. M., Johnson, L. R., Woodson, K. M., Henkin, A., & Dee, J. R. (2010). Empowering Women Faculty in STEM Fields: An Examination of Historically Black Colleges and Universities. *Journal of Women and Minorities in Science and Engineering*, *16*(4), 319–341. doi:10.1615/JWomenMinorScienEng.v16.i4.30

Mack, K. M., Rankins, C., & Woodson, K. (2013). From graduate school to the STEM Workforce: An entropic approach to career identity development for STEM women of color. *New Directions for Higher Education*, *163*(163), 23–34. doi:10.1002/he.20062

Mack, K. M., Rankins, C. M., & Winston, C. E. (2011). Black women faculty at Historically Black Colleges and Universities: Perspectives for a national imperative. *Diversity in Higher Education*, *11*, 149–164. doi:10.1108/S1479-3644(2011)0000011012

McGee, E. O. (2016). Devalued Black and Latino racial identities: A byproduct of college STEM success? *American Educational Research Journal*, *53*(6), 1626–1662. doi:10.3102/0002831216676572

Mickelson, R. A., Parker, A., Stearns, E., Moller, S., & Dancy, M. (2015). Family matters: familial support and African American female success. In Contemporary African American families: achievements, challenges, and empowerment strategies in the 21st century. New York: Routledge-Taylor & Francis.

Milgram, D. (2011). How to recruit women and girls to the science, technology, engineering, and math (STEM) classroom. *Technology and Engineering Teacher, 71*(3), 4–11.

Mirza, H. (2009). Plotting a history: Black and colonial feminisms in "new times.". *Race, Ethnicity and Education, 12*(1), 1–10. doi:10.1080/13613320802650899

Muller, C., Blake-Beard, S., Barsion, S., & Wotipka, C. (2012). Learning from the experiences of women of color in mentornet's one-on-one program. *Journal of Women and Minorities in Science and Engineering, 18*(4), 315–335. doi:10.1615/JWomenMinorScienEng.2013004006

Murrell, A. J., Forte-Trammel, S., & Bing, D. (2008). *Intelligent mentoring: How IBM creates Value through people, knowledge and Relationships*. Boston, MA: Pearson Education, Inc.

National Science Foundation, Division of Science Resources Statistics. (2006). S&E Degrees by Race/Ethnicities of Recipients: 1995-2004. NSF 07-309.

Nelson, R. (2018) Still Separate, Still Unequal: The Role of Black Women and Girls in the Legacy of *Brown v. Board of Education. Education & Training, AAUW.* Retrieved from: https://www.aauw.org/2018/05/17/still-separate-still-unequal/

Ogbu, J. U. (1994). Racial stratification in the United States: Why inequality persists. *Teachers College Record, 96*, 264–298.

Osanloo, A., Boske, C., & Newcomb, W. (2016). Deconstructing macroaggressions, microaggressions, and structural racism in education: Developing a conceptual model for the intersection of social justice practice and intercultural Education. *International Journal of Organizational Theory and Development, 4*(1).

Pascarella, E., & Terenzini, P. (2005). How college affects students: Volume 2, a third decade of research. San Francisco: Jossey-Bass.

Peffley, M., Hurwitz, J., & Sniderman, P. M. (1997). Racial Stereotypes and Whites' Political Views of Blacks in the Context of Welfare and Crime. *American Journal of Political Science, 41*(1), 30–60.

Perez, T., Cromley, J. G., & Kaplan, A. (2014). The role of identity development, values, and costs in college STEM retention. *Journal of Educational Psychology, 106*(1), 315–329. doi:10.1037/a0034027

Perna, L. W. (2001). The contribution of historically Black colleges and universities to the preparation of African Americans for faculty careers. *Research in Higher Education, 42*(3), 267–294. doi:10.1023/A:1018869922415

Pinder, P. J. (2008). *The "Black girl turn" in research on gender and science education: Toward exploring and understanding the early experiences of Black females.* Retrieved from ERIC database. (ED499925)

Powell, L. (1990). Factors associated with the underrepresentation of African Americans in mathematics and science. *The Journal of Negro Education, 59*(3), 292–298. doi:10.2307/2295564

Prime, G. M. (Ed.). (2018). *Centering Race in the STEM Education of African American K–12 Learners*. Bern, Switzerland: Peter Lang US. Retrieved Jan 8, 2020, from https://www.peterlang.com/view/title/67796

Purdie-Vaughns, V., Steele, C., Davies, P., Ditlmann, R., & Randall Crosby, J. (2008). Social Identity Contingencies: How Diversity Cues Signal Threat or Safety for African Americans in Mainstream Institutions. *Journal of Personality and Social Psychology, 94*(4), 615–630. doi:10.1037/0022-3514.94.4.615 PMID:18361675

Rainey, K., Dancy, M., Mickelson, R., Stearns, E., & Moller, S. (2018). Race and gender differences in how sense of belonging influences decisions to major in STEM. *International Journal of STEM Education, 5*(10), 10. doi:10.118640594-018-0115-6 PMID:30631700

Ricks, S. A. (2014). Falling through the Cracks: Black Girls and Education. *Interdisciplinary Journal of Teaching and Learning, 1*(4), 10–21.

Riegle-Crumb, C., Moore, C., & Ramos-Wada, A. (2010). Who wants to have a career in science or math? Exploring adolescents' future aspirations by gender and race/ethnicity. *Science Education, 95*(3), 458–476. doi:10.1002ce.20431

Rowley, S. J., Sellers, R. M., Chavous, T. M., & Smith, M. A. (1998). The relationship between racial identity and self-esteem in African American college and high school students. *Journal of Personality and Social Psychology, 74*(3), 715–724. doi:10.1037/0022-3514.74.3.715 PMID:9523414

Sellers, R. M., Chavous, T. M., & Cooke, D. Y. (1998). Racial ideology and racial centrality as predictors of African American college students' academic performance. *The Journal of Black Psychology, 24*(1), 8–27. doi:10.1177/00957984980241002

Settles, I. H., Cortina, L. M., Stewart, A. J., & Malley, J. (2007). Voice matters: Buffering the impact of a negative climate for women in science. *Psychology of Women Quarterly, 31*(3), 270–281. doi:10.1111/j.1471-6402.2007.00370.x

Singleton, G. E. (2015). Courageous conversations about race: a field guide for achieving equity in schools. Thousand Oaks, CA: Corwin, A SAGE Company.

Smith, J. L., Lewis, K. L., Hawthorne, L., & Hodges, S. D. (2013). When trying hard isn't natural: Women's belonging with and motivation for male-dominated STEM fields as a function of effort expenditure concerns. *Personality and Social Psychology Bulletin, 39*(2), 131–143. doi:10.1177/0146167212468332 PMID:23187722

Steele, C. M. (1997). A threat in the air: How stereotypes shape intellectual identity and performance. *The American Psychologist, 52*(6), 613–619. doi:10.1037/0003-066X.52.6.613 PMID:9174398

Steele, C. M., & Aronson, J. (1995). Stereotype threat and the intellectual test performance of African Americans. *Journal of Personality and Social Psychology, 69*(5), 797–811. doi:10.1037/0022-3514.69.5.797 PMID:7473032

Steele, C. M., Spencer, S. J., & Aronson, J. (2002). Contending with images of one's group: The psychology of stereotype and social identity threat. In M. P. Zanna (Ed.), Advances in experimental social psychology (34th ed., pp. 379–440). San Diego: Academic Press.

Sundermier, A. (2018) Rewiring STEM education. *Symmetry Magazine.* Retrieved from https://www.symmetrymagazine.org/article/rewiring-stem-education

Syed, M., Azmitia, M., & Cooper, C. R. (2011). Identity and academic success among underrepresented ethnic minorities: An interdisciplinary review and integration. *The Journal of Social Issues, 67*(3), 442–468. doi:10.1111/j.1540-4560.2011.01709.x

Terenzini, P., & Wright, T. (1987). Influences on students' academic growth during four years of college. *Research in Higher Education, 26*(2), 161–179. doi:10.1007/BF00992027

Thiry, H., Laursen, S. L., & Hunter, A. B. (2011). What experiences help students become scientists? A comparative study of research and other sources of personal and professional gains for STEM undergraduates. *Journal of Higher Education, 82*(4), 357.

Thompson, P. (2008). On firm foundations: African American Black college graduates and their doctoral student development in the Ivy League. In M. Gasman & C. L. Tudico (Eds.), *Historically Black Colleges and Universities: Triumphs, Troubles, and Taboos*. New York: Palgrave Macmillan. doi:10.1057/9780230617261_3

Tinto, V. (1993). *Leaving college: Rethinking the causes and cures of student attrition*. Chicago: The University of Chicago Press.

Toppo, G., & Nichols, M. (2017). *Decades after civil rights gains, black teachers a rarity in public schools.* https://www.usatoday.com/story/news/nation-now/2017/02/01/decades-after-civil-rights gains-black-teachers-rarity-public-schools/96721684

Twin Cities Public Television. (2013). *The SciGirls Seven, Proven Strategies for Engaging Girls in STEM*. Retrieved from http://www.scigirlsconnect.org/wpcontent/uploads/2016/07/SciGirls_Seven_2Page.pdf

Van Eps, M., Cooke, M., Creedy, D., & Walker, R. (2006). Students' evaluation of a year-long mentorship program: A quality improvement initiative. *Nurse Education Today*, *26*(6), 519–524. doi:10.1016/j.nedt.2006.01.009 PMID:16540212

Van Langen, A., & Dekkers, H. (2005). Cross-national differences in participating in tertiary science, technology, engineering and mathematics education. *Comparative Education*, *41*(3), 329–350. doi:10.1080/03050060500211708

Wenglinsky, H. (1997). Students at Historically Black Colleges and Universities: Their aspirations & accomplishments. Policy Information Report. Princeton, NJ: Educational Testing Service.

Will, M. (2018) *There Are Many More Female STEM Teachers Now Than 20 Years Ago.* http://blogs.edweek.org/edweek/teacherbeat/2018/10/who_is_the_average_stem_teacher.html

Wilson, A., Sanner, S., & McAllister, L. (2010). An evaluation study of a mentoring program to increase the diversity of the nursing workforce. *Journal of Cultural Diversity*, *17*, 144–150. PMID:22303649

Yeager, L. (2000). *Mentoring: Perceived benefits and challenges for a university setting*. (Dissertation). Spalding University, Louisville, KY.

Zirkel, S. (2002). Is there a place for me? Role models and academic identity among White students and students of color. *Teachers College Record*, *104*(2), 357–376. doi:10.1111/1467-9620.00166

Zirkel, S. (2004b). What will you think of me? Social integration and educational achievement among White students and students of color in elementary and junior high school. *The Journal of Social Issues*, *60*, 57–74. doi:10.1111/j.0022-4537.2004.00099.x

Zirkel, S., & Cantor, N. (2004). 50 Years After Brown v. Board of Education: The Promise and Challenge of Multicultural Education. *The Journal of Social Issues*, *60*(1), 1–15. doi:10.1111/j.0022-4537.2004.00096.x

Chapter 7
Rediscovering Our Original Selves:
What Did We Leave Behind (and Pick Up) on the Journey to "Success"?

Christine Sharon Grant
North Carolina State University, USA

ABSTRACT

STEM careers are complex, requiring the management, monitoring, and implementation of several different technical and interpersonal interactions. For women of color (WOC), the journey is complex as they are impacted by issues of bias, marginalization, and unconscious bias. Finding opportunities to reflect, rejuvenate, and regenerate requires strategic planning to explore the positive and negative elements in your STEM career box. In this work, the author reflects on a 30 year STEM career, challenging the reader to explore their own career trajectory in the context of a conceptualized STEM success cycle.

INTRODUCTION

Our Reality, Our Experience, Realism Defined

For many of us, we can clearly define the moments in which we made a decision to pursue a career in STEM. For some it was a science experiment in a middle school afterschool enrichment program, for others it was the pre-college camp to broaden participation in STEM. Some of us "fell" into a career in STEM based on

DOI: 10.4018/978-1-7998-4858-5.ch007

the information provided by an academic mentor or a community advocate. Another segment of us may have always known that STEM was an interest, and perhaps there was a pivot to do it full time in our pursuit of a college degree. You know, the proverbial switch from pre-law to one of the traditional STEM fields may have occurred during freshman year in college. Someone may have told us about the rigor associated with the journey, and the "prize" career at the end of the tunnel. We may have even been following in the footsteps of a famous STEM professional. One can be fairly certain, however, that we were not prepared for the sometimes soul-crushing elements that we would face as a Woman of Color in pursuit of a STEM career. The looks, the questions, the assumptions and even the dismissive behaviors that some of us seal away in a closed container (or even a journal). Hiding incidents and masking the impact on our psyche and our STEM soul in the depths of our consciousness may actually be a common practice.

The interesting thing is while our experiences may differ, our initial draw to STEM was likely very similar to that of our colleagues. We loved the quest for knowledge, the open-ended questions and the discovery of answers to questions that could help humanity. It all started with a spark. A quest to learn scientific principles and apply that information to the solution of complex problems was that spark. As a pre-college student in summer Minority Introduction to Engineering (MITE) programs, we may have been taught by dedicated university professors, engaging graduate students and counseled by undergraduate students who wanted us to join their ranks. We may have known that our numbers in the field were low, but our imaginations were provoked to explore the possibilities to become engineers and technological professionals. We adjusted our goals upward and learned the required milestones to be "successful" in STEM. We were taught (and we believed) that one could adjust and create one's own realism and manage the elements required for success. Given the working model for this nuanced reality, we vigorously pursued the roadmap provided for STEM success. We worked hard without taking long reflective breaks.

During a unique, "sabbatical like" break, the author took a long overdue look at her own 30-year STEM journey juxtaposed with the terms and concepts associated with women in STEM. Not unlike many women in STEM, she experienced her career through the limited lenses of close colleagues, friends and family and perhaps the occasional mentor/coach who had a similar social identity. As a woman of color, the author could harken back to the stories from her parents about the challenges they faced as African Americans in the era from the 1940's to the 1970's; she used those narratives as a backdrop to explain some elements of what she was experiencing as a STEM professional. It became the context of her experience and her reality. There was, however, something missing.

The author had the good fortune of being part of a series of networks, communities and initiatives as a backdrop to put her experiences as a "lonely only" WOC in the proper context. Better yet, there was a community of practice that was defining, studying and partnering for positive change. These included conferences for Women of Color (WOC) in STEM, National Science Foundation ADVANCE programs and myriad workshops held by the National Academies, the American Society of Engineering Education (ASEE), Institute for Women's Policy Research (IWPR), and the American Association for the Advancement of Science (AAAS). These initiatives have a plethora of reports that provide excellent definitions of the challenges and opportunities for change. Their recommendations and strategies for implementation of promising practices provide resources for both organizations and individuals to make impact the success of WOC in STEM.

Gradually the author discovered that her experiences could be put it into the social context of race, gender, equity and inclusion with a broadened set of concepts. For example, unconscious bias, micro-aggression, double-bind and imposter syndrome became concepts that formed a framework to explain and more importantly, to understand her STEM journey. There are certainly more opportunities for WOC in STEM to interact, share their realities, and learn about these concepts at themed conferences, in online spaces, and in professional spaces (e.g., workplace, society meetings) than when she entered college in 1980.

After years of exposure to the concepts of unconscious bias, micro-aggressions, inclusivity, and stereotype threat (to name a few), the author reviewed the words penned over 20 years ago in her journals. As she reviewed the harsh reality of a set of exchanges with leadership and colleagues, she realized that there were several elements of her journey that she had suppressed. She recalled some of the feelings, but she had blotted out the details of the exchanges. After this review, her reality was "redefined". She developed a list of elements for what she defines as a STEM Career Box...reflecting the range of experiences, emotions and just as important, the aforementioned theoretical concepts (e.g., unconscious bias) that mapped onto the narrative of her life.

It is also critical, the author observed, that STEM professionals and in particular, WOC in STEM, should not wait 30 years to conduct a self-reflective review of their experiences. The STEM Success Cycle and the STEM Career Box were developed as a result of her experiences and informed by the literature.

BACKGROUND

There is a wealth of literature (including core chapters in this book) that extracts the unique elements that are faced by WOC in STEM in a range of workplace environments. In the seminal 1975 paper entitled, "The Double Bind: The Price of Being a Minority Woman in Science" and the 2011 updated version (L. E. Malcom & Malcom, 2011; S. M. Malcom et al., 1976), a group of women STEM professionals convened to discuss the challenges, opportunities and path forward to fully engage Women of Color (WOC) in successful pursuit of a STEM career. The summary indicated that:

These thirty minority women scientists from the Native American, Black, Mexican-American and Puerto Rican communities have found their careers in science, engineering and medicine to be exciting, demanding and rewarding. Their extraordinary resilience and commitment have overcome the obstacles that have discouraged many others. While reluctant to concede their own specialness, the participants in this conference were unanimous in their determination to change things so that other able minority women need not be so very special to have careers in science. (S. M. Malcom et al., 1976)

More recently, reports by the National Society of Black Engineers (NSBE) and the Society of Women Engineers (SWE) explored critical issues (with recommendations) at the intersection of race and gender for WOC engineers. The NSBE report entitled, "Ignored Potential: A Collaborative Roadmap for Increasing African American Women in Engineering" indicated that:

"Researchers cite several systemic factors that work against the participation of African American women in engineering ((Foor et al., 2007; Obiomon et al., 2007). Lack of visible role models in engineering, stereotype threat, biculturalism, tokenism, feelings of isolation, and pay inequities in the engineering workforce are all factors at play. Black women in engineering must put in additional effort to overcome these unique challenges that occur at the intersection of racial and gender inequality (Fletcher et al., 2017)."

A Society of Women Engineers (SWE) study entitled, "Women of Color in the Engineering Workplace: Early Career Aspirations, Challenges, and Success Strategies" (Rincon & Yates, 2018) indicated that: *"Stereotype threat may subconsciously discourage women of color from pursuing engineering careers. When a demographic with associated negative stereotypes becomes salient, individuals who fit that demographic will often experience poor performance due to the fear of confirming those stereotypes (Steele & Aronson, 1995)."*

The SWE study also indicated the following challenges expressed by women in the study: (i) lack of role models for minority female engineers, (ii) disillusionment regarding level of impact they would have as engineers, (iii) gender and racial biases and stereotypes experienced in the workplace, (iv) dissatisfaction with salary and benefits, (v) unfair performance evaluations and lack of honest feedback, and (vi) difficulty obtaining professional development through employers. For the author, these challenges resonated with her lived experiences as a WOC in STEM.

An NSF supported report by the American Association of University Women (AAUW) entitled, "Why so Few? Women in Science, Technology, Engineering, and Mathematics" addressed elements of bias saying that: *"If women and men in science and engineering know that this bias exists, they can work to interrupt the unconscious thought processes that lead to it. It may also help women specifically to know that if they encounter social disapproval in their role as a computer scientist or physicist, it is likely not personal and there are ways to counteract it (Hill et al., 2010)."*

There are several comprehensive scholarly works that articulate the issues and solutions for STEM WOC at different points in their career. In the paper, "Linking intersectional invisibility and hypervisibility to experiences of microaggressions among graduate women of color in STEM", researchers (Wilkins-Yel et al., 2019) reviewed the salient terms and "*explored recommendations for addressing such experiences of microaggressions and intersectional invisibility*". Their work *"illuminated the enormous time and energy that is needed to survive STEM environments that is above and beyond doing STEM."* They recommended *"ongoing cultural competency training for the faculty, staff, and students"* and support systems in the form of same race and gender networks.

In another work (Ong et al., 2018) entitled, "Counterspaces for women of color in STEM higher education: Marginal and central spaces for persistence and success", the authors focused on persistence and promising practices for a groups of women in their study. This study indicated that, "*for the sake of retaining women of color in STEM, more attention must be paid to the safe social spaces, or counterspaces, which offer support and enhance feelings of belonging in STEM*". Another aspect of the work indicated that: *"...peer relationships, mentoring relationships, and STEM diversity conferences—all counterspaces that involved participation by members at multiple stages (e.g., undergraduates, graduate students, and faculty)—enabled bridge-building across levels of differential power."* The importance of counterspaces is echoed in the work entitled, "Microaggressions in Engineering Education: Climate for Asian, Latina and White Women" which emphasized "*the gendered and racialized nuances of inequities in engineering and how women continuously construct counterspaces to adapt within this segregated academic environment (Camacho & Lord, 2011).*"

The reality is that people can attempt to impact your experiences in a negative manner, influence your history (or historical narrative) and change your realism. The ability to rediscover your original self, reset the roadmap, revitalize your game plan, regain your confidence and restore your resilience requires a degree of reflection at various stages in your roadmap. It's almost like one needs a course correction as one pursues the path and potentially the passion associated with a STEM career. This chapter will engage you in the aforementioned elements of this process while framing the core issues through the experiential perspectives of the author. The chapter will also incorporate "sanitized" excerpts from the author's journals, providing real-life thoughts during a thirty-year STEM journey. The core concepts and examples are applicable to WOC in industry and academia at all stages of their careers. It is also critical, the author observes, that STEM professionals and in particular, WOC in STEM, should not wait 30 years to conduct a self-reflective review of their experiences. The STEM Success Cycle and the STEM Career Box were developed as a result of her experiences and informed by the literature. The overarching goal is to compel you to look at your own journey and identify, create and execute a revitalized plan to fully experience your "moments of success" while putting the challenging experiences in the proper perspective.

What Color Is Your Lunch Box?

Did you ever have a lunch box? Perhaps it was a metal box with hinges or maybe it was a vinyl bag with a cool handle. The author remembers each year as part of the pre-college back to school supply acquisition, she would get a new lunchbox with the cool cartoon characters on it. When times got hard economically (in the 1970's), her parents made her keep the same lunch box for consecutive years. If you took your lunch to school or work, early on there may have been a family member or guardian who packed your bag to take to school. Maybe you got a note encouraging you, or there was a special snack with tasty leftovers from the weekend before. As responsible adults, we often find ourselves packing our own lunches; making decisions about healthy and not so healthy food and snacks. Have you ever considered purchasing a new, "oh so colorful" lunch box with really cool handles on it? Perhaps you realized that you actually had several perfectly functional boxes at home in the kitchen closet. Furthermore, you may have realized that it was not important what the box looked like if you were not going to pay attention to the contents inside.

The decision to not buy another lunch box and instead pay attention to what is in the one being carried now is analogous to how we may be treating elements of our STEM careers, in what we will call a STEM Career Box. As your careers progresses, of course you need to pay attention to what the outer box presents to ourselves and to others. There should always be an external degree of professionalism and an expectation of excellence that defines your STEM success. More importantly, however, it is the internal contents that provide the required STEM career nutrients for your educational and career pursuits.

As we endeavor to "discover our original selves" it is important to understand that while we have picked up empowering elements, we must also seek to discard detrimental items from our STEM Career Box. Whether you are at the beginning of your STEM career, reflecting on a mid-career juncture or preparing to enter a senior phase of your STEM journey, challenge yourself to ask the following two questions at every stage in your career:

1. What elements did I pick up on my journey to success?
2. What elements now need to be discarded as I move to the next stage of my journey?

Furthermore, challenge yourself to *get back to your true self* after years of "adapting, fitting in, and making professional linkages" in a manner that is authentic, empowering and professional.

Nice Lunchbox: But What Is Inside?

You should care about the contents of your professional STEM Career "lunch" box and more importantly how they impact you when you consume the contents. The reality is that oftentimes no one cares how or what you carry. After an extensive review of years of personal journals, the author began to see a trend in the negative and positive elements of her own career in STEM. She grouped them together an developed a list of 14 elements.

Seven potentially positive elements of the STEM Career Box are listed in Table 1. This is not an exhaustive list. Perhaps there are elements in your STEM Career Box that are variations on the seven elements, they may even be different altogether. The degree or condition of the elements can contribute positively or negatively to our STEM career. For example, the absence of suitable amounts of respect, wise counsel, balance, or wellness can adversely impact you in your STEM career. Interestingly, just like a piece of moldy fruit can eventually spoil the contents of a lunchbox, so can negative elements adversely impact the positive elements of the STEM Career Box. In our professional box M.O.L.D. can Make Our Lunchbox Decline. Some elements that can be distractions and work to our detriment are listed in Table 1.

A closer look at these 14 elements reveals that some of them represent: how we feel about ourselves, how we feel about other people, and institutional infrastructure issues. A critical aspect of this analogy is that we cannot control other people and what they put in our Career box, nor can we easily control the environment, culture or personal attributes of the people in our lives. We can, however, work individually and collectively to enlighten, challenge and change organizations to reduce the toxicity that causes M.O.L.D.

Table 1. 14 Potential Positive and Negative Elements in Your STEM Career Box

Potential Positive Elements	Potential Negative Elements
1. Wise counsel from coaches, advocates, mentors, etc.	1. Stress, frustration and pent up hostility
2. R.E.S.P.E.C.T.	2. Regret, Shame and feelings of unworthiness
3. Positive interpersonal interactions/negotiations	3. Bias, micro and macro aggressions
4. Reflection, Rejuvenation, Regeneration and Revitalization	4. Inequitable allocation of resources and opportunities
5. Balance, Transitions and Shifting needs	5. Long term anger, jealousy
6. Shifting definitions of "success"	6. Unhealthy competition
7. Wellness, well-being and positive psychology	7. Unfinished business

Later in this chapter, we challenge you to identify the elements of your own STEM Career Box, create an understanding of how they relate to where you are in your journey and execute a process of removing the elements that are toxic to your journey. At the onset, we must understand that the STEM professional environment is challenging. It is a challenge to men and women from all backgrounds. The global competition one has to "win" to both survive and thrive in a STEM academic program is unprecedented in nature. Attaining a degree in a STEM discipline is intellectually challenging and intellectually rewarding at the same time. These challenges and opportunities are also part of the STEM Career. An enlightened understanding of the elements in your Career Box facilitates your success. Furthermore, managing these elements are critical to your actual success, both real and perceived.

STEM Success Cycle

Success can be defined in several different ways. The traditional definition in the online dictionary says that success is (*Definition of SUCCESS2*, n.d.):

- the accomplishment of an aim or purpose.
- the attainment of popularity or profit.
- a person or thing that achieves desired aims or attains prosperity.

The Merriam-Webster dictionary defines the noun success as (*Definition of SUCCESS*, n.d.)

- degree or measure of succeeding
- favorable or desired outcome, *also*: the attainment of wealth, favor, or eminence

- one that succeeds

The act of being successful is described as (*Definition of SUCCESSFUL*, n.d.)

- resulting or terminating in success
- gaining or having gained success

One's personal definition of STEM success can range from accomplishing an education goal to completing a technical project to empowering others to excel in STEM. If we gave ten people the task of defining STEM success, they would likely have themes that were similar, but their priorities would differ. There are three phases to what we will call the "STEM Success Cycle" (see Figure 1), they are as follows: (1) Learning the rules, (2) Paying your dues and (3) Making it you…

Figure 1. STEM Success Cycle

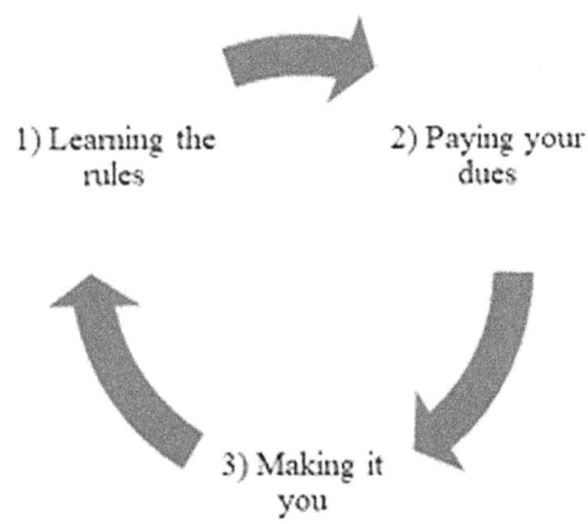

It is important for you to understand: (i) what these three phases are, (ii) how one moves from one phase to another, (iii) how to learn from the STEM Success Cycles of others and (iv) how to clearly articulate your needs in these phases to positively impact your professional journey. These three phases can happen in rapid succession or in a cycle that manifests itself in varying timelines with respect to different projects in the realms of leadership, research, teaching, and other STEM related activities. Similar to the stickers that we may have put on the outside of our lunchbox, we can think of the STEM Success Cycle as an emblem placed on the outside of our STEM Career Box. Let's explore how the three stages may manifest in general and then explore further with specific examples from the author's career.

Learning the Rules

This phase includes the process of someone or something providing you with the written (and the unwritten) rules for the job at hand. This could be in the form of a training manual, personal interactions, a class/orientation or a longer-term training initiative within your organization. While there are several mechanisms (e.g., verbal, written, online, face-to-face) that can be utilized to provide this formal training, it is important to know your learning style, so that you can maximize the impact. In some instances, one may find it necessary to request as different approach to receiving information. For example, asynchronous receipt of information via an "on-demand" media portal is a great way for those who have attention issues or demands that do not enable long periods of uninterrupted receipt of content.

The person who delivers the "rules" could be a direct supervisor, an assigned mentor, a senior member of your team or an HR professional external to your group. There could be wide variations in the effectiveness of the delivery of the rules. In rare instances, the hidden agenda of the delivery person may impede your ability to get honest and true information on the norms and expectations associated with the rules. There could also be "hidden or unwritten rules" associated with the way things are done in your organization.

Paying Your Dues

This is one of the more challenging elements of the STEM Success Cycle. It is in this stage, one is becoming or is well integrated into the fabric of the organization, pressing upward and moving/climbing the ladder of success. The hardest part about this stage is wondering if (and feeling like) the "rules of engagement" are the same for you as they are for persons who do not look like you. This is the stage that you work diligently to "make it work", assuming that the metrics for success (and the associated celebrations) will be equitable as you work your career plan. In a perfect scenario, you will have equivalent resources, and of course the same opportunities to make the entire thing work for you and the organization. Everyone's STEM journey is an individual pathway, strewn with pebbles, obstacles, potholes and turns that are unique to each person. The question is, "When you stumble and when you fall, what resources are there to pick you up and place you at the same point or ahead of where you fell?" Falling backwards or getting up and being pushed backwards can be unexpected consequences in the STEM journey. This is another critical point at which mentors, coaches, allies, sponsors and advocates can play a pivotal role in success.

Making It You

Putting your stamp on something and making it your own is a powerful thing. The ability to make something yours depends on how the power structures are set up, how long you are willing to wait to own something and the systems in place (or emerging) to empower you to make it yours. This concept applies to many different aspects of a STEM career. It could apply to something physical or related to personnel, like a laboratory, office, direct reports or resources for travel. It could also be in reference to: having ownership of the implementation of ideas, the freedom to engage with new colleagues and collaborators, or your ability to make decisions about directions of teams of personnel. Finally, one's ability to design and implement innovations that change the infrastructure by changing the fabric of an organization can come from a leadership role or being a member of a managing group (e.g., board of directors, advisory roundtable).

STEM Success Cycle Examples

The following two scenarios, from the author's experience, highlight aspects of the STEM Success Cycle and how they could manifest. More important than these two examples are your own reflections on the phases in your own STEM career. The author recalls the following experience:

My colleagues told me that the meeting to assign graduate students to faculty advisors occurred in a general faculty meeting late in the fall semester. When I arrived to the meeting and saw the assignment, I was a bit puzzled at the student I had been given as my first student as a new faculty member. The faculty around the table were urging me to seriously consider that student, after all a mediocre student was better than no student at all. It was after the formal meeting that I found out that there were several 'meetings before the meeting' in which colleagues had caucused to negotiate who they wanted as student research advisees. (By the way I did not take the student as an advisee)

A few years after the above incident, the department implemented revised rules that were a bit more transparent for both students and faculty alike. While there is always some level of murkiness associated with the allocation of resources, the more equitably things are done, the better it is for all involved. Learning how to attract research mentees, collaborate with colleagues in research and educational ventures, and all the associated "stuff" was the paying dues component. Fast forward 20 years, as the Associate Dean of Faculty Advancement in the College of Engineering, the author has myriad opportunities to develop and implement initiatives to promote faculty success, while facilitating transparent transfer of information to new and established faculty. It should be pointed out that not all "making it you" implementations have a 20-year timeline. In some instances, one can implement something that is revolutionary, impactful, unique to your own experience and your method of delivery in a short time.

In a different scenario, that was still related to research personnel, the author recounts her experiences as the co-lead of a large program that provided research mentoring for undergraduate students. The Research Experience for Undergraduates (REU) program (an NSF funded program) provides financial support for approximately 20-30 students each summer to perform research with different faculty. The faculty coordinators of the program also supervised a program manager in addition to crafting career development (e.g., how to apply to graduate school) and community building activities for students over the course of 8-10 weeks in the summer. In the pursuit of academic tenure, early career faculty are evaluated in the realms of research scholarship, teaching and extension. In many cases, the acquisition of research funding is an important element in the evaluations; in some instances, it is considered more important than teaching.

Undergraduate research and the mentoring of undergraduate students is an important component of training the next generation of STEM professionals. There is, however, not a 'standard' mechanism for early career faculty to get 'credit' for this activity. A colleague and I partnered to write a successful NSF grant and led an NSF REU Site program for students to come from other institutions to our university to do work with a range of faculty across the engineering disciplines. We led a $500K+ program for upwards of a hundred undergraduates who impacted the research of dozens of faculty/postdoc colleagues in our college. During my tenure review, there was discussion as to whether this was research or teaching.

In this instance, it was clear that the long-term faculty career impact of mentoring undergraduates was not perceived as a core critical task for new faculty, this was the "anecdotal rule". We paid our dues and wrote the proposal, coordinated the program (and associated personnel) for several years, making it our own. It was important to us and we converted this important activity into a funded project that impacted students and faculty. While at the end of the day, there was still some question about whether this should "count" as research funding during the promotion process, it was an initiative that could not be ignored.

It's Your Turn (Part 1: STEM Success Cycle)

Past, present and future projects, situations and initiatives are at the heart of STEM professional development. Using the three phases of the STEM Success Cycle (see Table 2): (i) identify a past situation when you completed all three steps in the cycle, summarize the three phases (ii) identify a present challenge in which you are either in stage 1 or stage 2 of the cycle, finally (iii) lay out a future scenario in which you need to strategize on how to accomplish all three aspects of the cycle. You can share your results with a mentor, trusted colleague or a friend. Talking through aspects of the scenarios can often provide different ways to improve the impact in each phase.

Table 2. Your STEM Success Cycle Exercise

Scenarios	Learning the Rules	Paying Your Dues	Making It You
Past			
Current			
Future			

One Woman's Journey and Her STEM Career Box

After 30 years as a member of a college of engineering faculty, the author decided to take a break. That's right… take a break. You may be wondering exactly what that means. How do STEM academics take a break? For that matter how do STEM professionals take a break? In the first place, how do you define "a break"? For the author, the break consisted of a 30-day period of reflection on the first 30 years of her career. She spent time reflecting on the contents of 20+ years of personal journals, rejuvenating herself by reconnecting with friends (and doing yoga), and regenerating/revitalizing the core elements of her STEM career portfolio. There were a few rules she had during this time, including: do not make any major decisions, reduce media consumption, get plenty of rest, explore the city and connect with work only if absolutely necessary. While the recounting of career reflections during the author's break may be interesting to you, the reader, it's more pertinent for you to consider applying the processes of the STEM Success Cycle and the STEM Career Box to your own career reflection and planning.

While many people may journal, they may not know the textbook definition of the journaling. In the article "Journaling Enters the 21st Century: The Use of Therapeutic Blogs in Counseling" (Lent, 2009) the authors indicate that, *"Journal writing is a cost-effective counseling intervention that can deepen self-reflection and facilitate the therapeutic change process... Daily journal writing provides a meditative period of self-reflection and introspection and deepens the process of self-analysis...Journal therapy is an effective means of providing focus and clarity regarding issues, concerns, conflicts, and confusions. Through journal therapy, the traditional diary is changed into a genuine, unique therapeutic method that offers cost-effective, holistic self-management."* Whether you journal daily, record thoughts on an electronic device or blog perspectives on your STEM journey, self-analysis of the contents can provide revelatory insights into the elements that you may have left behind or picked up on your journey to success.

Let's start at the beginning. Almost 40 years ago the author began her career at Georgia Institute of Technology as a graduate student in chemical engineering. She joined the engineering faculty of her current institution as the first African-American woman faculty member and the only one in the college of engineering for 16 years. Thirty years later there are still less than five AA women out of almost 400 engineering faculty. It's important to note that the overall experience has been a rich one, filled with positive elements of collegiality, collaboration and intellectual challenge. While a glorified or a gory account of the journey will not be presented here, the author will attempt to provide a balanced perspective of the journey as a STEM professional. The highs and the lows are not unique, you likely have also experienced the highs and lows over the course of your career. Perhaps you are just starting your career and are wondering how to manage the professional and the personal elements of the journey. There are aspects of the experiences of STEM professionals that transcend the STEM career sector (i.e., industry, academia, business, etc.). More specifically, there are experiences that Women of Color in STEM have that form a common thread across the disciplines.

Whenever one takes a break from a very intense activity, there is a time to be reflective and wax philosophical about different aspects of the activity. In this case, the journey can be described with the title "Thirty Years an Academic". That is, for thirty years the author has impacted the institution and for thirty years that the institution has impacted her. It has been a key aspect in shaping elements in her own STEM Career Box. Fortunately, there are personal journals that have chronicled the journey starting over 30 years ago. This chapter will incorporate aspects from journals that could be considered "mid-career", after approximately 15 years as a faculty member. In this regard, the entries don't represent very early career frustrations, nor do they reflect on the long-term seasoned career perspectives. While there are not entries for each of the 14 STEM Career Box elements, the ones that are included have changed some details to protect the innocent… or in some cases the "not so innocent" persons and events in the author's STEM journey. Interestingly, the following 14 elements (positive and negative) can be distractions and work to our detriment;

the key is to understand them and develop methods to address them. The goal in the next section is to identify the issues and point the reader to explore their own experiences (potentially via writing) – without a major focus on the authors journey. In fact, the examples given may even be the opposite of your own experiences. The objective is to get you to explore your personal narrative (e.g., journals, blogposts, other writings) in the context of the 14 elements of the STEM Career Box.

Wise Counsel from Coaches, Advocates, Mentors, Etc.

There is a wealth of information available on the importance of mentors, coaches, allies and advocates in the career of a STEM professional. There are also a number of conferences and initiatives dedicated to the professional development of WOC in STEM (*ACS Women Chemists of Color Program*, n.d.; *ACS Women Chemists of Color Program*, n.d.; *Women Chemists Committee*, n.d.; Carrigan et al., 2018.; Horner-Devine, 2019.; Nave & Williams, 2018). In one instance, a series of conference sessions and workshops by the Women of Color in the American Chemical Society (ACS) resulted in a book entitled, "Growing Diverse STEM Communities: Methodology, Impact and Evidence" focused on the narratives from different STEM segments. In the chapter entitled, "Mentors, Mentors Everywhere: Weaving Informal and Formal Mentoring into a Robust Chemical Sciences Mentoring Quilt" the author indicated that,

Informal and formal mentoring manifest in different ways for women and underrepresented minorities in the chemical sciences. This is particularly true for Women of Color (WOC) who occupy a unique place at the intersection of race and gender in STEM fields. While there is still more work to be done to achieve parity, equity, and inclusion for diverse groups in the chemical sciences, an intentional, informed, and bidirectional pursuit of mentoring can result in robust, long-lasting mentoring teams. (C. S. Grant, 2019)

The ACS Chapter chronicles in detail who the author's mentors were in the chemical sciences and what sectors they worked in (i.e., industry, academia, national labs, government). There are also narratives from other WOC in the chemical sciences related to mentoring in the ACS chapter. The exercise of going back to high school to identify early mentors and drawing the mentoring thread through current mentors and coaches over forty years later was a powerful exercise that is highly recommended for STEM professionals. It's been their wisdom and the experiential elements of the author's career that provide the backdrop to putting her elements in the proper perspective.

Of course, there are non-STEM advocates in our lives who are oftentimes just as impactful on our career success. In the case of the author, she had parents who were both educators (one in STEM and the other in the arts). She says,

I am thankful! For parents who both care and support me – for people that love and protect me and for wisdom to work hard. (Grant JE, 2/3/05)

The author views active utilization of coaches, mentors, allies and advocates as key elements that she picked up along the way.

R.E.S.P.E.C.T.

There are varying levels of respect and disrespect that STEM professionals all experience in their lives. In some instances, one cannot control the negative elements that are present. Yes, one should expect respect as a default as a STEM professional in both work and social situations. However, you need to be prepared when you don't get respect and not be surprised. In a work entitled, "Navigate, Balance, and Retain: Developing Success in the Mid-Career for Female STEM Faculty (McCord et al., n.d.)", the authors discussed perceptions of respect and inclusivity for mid-career women STEM faculty. They found that most respondents (male and female) in the study indicated that they were treated with respect by their colleagues. However, a closer look at the differences by gender indicated that: *"STEM females were less likely then their male counterparts to feel like full and equal participations in departmental problem-solving and decision-making."*

For the author it was a surprise how respect could manifest itself in public venues in front of colleagues. She recalls the aftermath of an experience in a monthly departmental faculty meeting:

I had a hard day yesterday – disrespect from Professor X in the faculty meeting and afterwards. I broke down crying in the stairwell afterwards – Professor Y consoled me. I went to his office to talk. He listened and gave some wise advice and counsel. (Note – the consoling faculty member was junior to me – yet in positive relationship with Professor X). (Grant JE, 11/16/02)

It also was not clear how one could, independent of others, demand respect from more senior colleagues:

Thoughts – Forgiveness for Professor X's disrespect of me and my abilities. There seems to be some resentment there and I wish I could address it. I need to set up an

appointment to have lunch with him to air my concerns and clear the air. Perhaps I'll take him out to lunch (I never did this...) (Grant JE, 11/19/02)

Years after the above two episodes, the author had the opportunity to attend an NSF ADVANCE conference at UNC Charlotte focused on promoting success for women STEM academics. The following journal entry indicates how she reflected on the above micro/macro-aggression event during her formal presentation on what should be expected from colleagues:

Don't disrespect and or insult me, speak to me when I walk into a room and not around me; engage me in light, caring, authentic conversations; ask my opinion.... (Grant JE, 2008)

While the above examples dominated the author's professional world for a season, there is actually a larger context that transcends the interactions of two or more colleagues. The concept of collegiality is one that ideally should be fostered by organizations, emanating from leadership and infused into the operations via training, policies and understanding of collegiality best practices. In the book "Facilitating a Collegial Department in Higher Education: Strategies for Success" (Cipriano & Cipriano, 2011), Robert Cipriano says that:

The most effective approach is one that addresses the concerns of incivility, non-collegiality and lack of respect in a systematic manner. There should be university-wide policies established that promote collegiality, civility, and respect campus-wide. Universally applied reward and reprimand systems should be established and consistently followed. It is true that what gets rewarded gets repeated. (Cipriano & Cipriano, 2011)

During the rediscovery of her original experiences, the author can now redefine the experiences in the context of aggressive behavior. The aforementioned episode in her career is described by Camacho and Lord:

"Microaggressions can be experienced by women in engineering in many forms: at the institutional level (such as through biases in the curriculum, structure and type of assignments, and even accepted canons of the knowledge base); at the interpersonal level (in the form of snubs, dismissive gestures, or seemingly-innocent comments that are perceived as hurtful, inappropriate, or insulting); and finally as jokes and humor that subtly deride women's place in engineering" ((Camacho & Lord, 2011).

When looking at one's own journey, identifying element of respect (or lack thereof) is the first step in addressing the issue. The second step is to learn the roles of organizational culture and individual personalities on the elements of respect. Finally seeking wise counsel from peers, leadership, mentors or HR can enable a clear path forward.

Interpersonal Interactions/Negotiations

From situational responses to strategic resolution of conflict, interpersonal interactions and the ability to negotiate requires a high degree of emotional intelligence. Joretha Johnson, an engineer and a retired executive leader in leading engineering organizations (who is also an African American woman), detailed the critical aspects of negotiating with emotional intelligence in a chapter in the book "Success Strategies from Women in STEM: A Portable Mentor". Her insights included the following: (i) negotiating is a process, (ii) a negotiation is a strategically focused conversation and (iii) negotiating is a balancing act (Johnson, 2015). The requisite five components of emotional intelligence developed by Goleman are: (i) self-awareness, (ii) self-regulation, (iii) motivation, (iv) empathy and (v) social skill (Goleman, 2014). All of these could be designated as additional tools in your box – to facilitate productive interpersonal interactions. The author recalls an unsuccessful negotiation and positive personal interactions in one entry.

Thankful that in spite of a really tough day on Monday – arguing with Professor X and Professor A about support for my post doc, today was a great day. I took my son to school, went to the gym – worked out at the YMCA. Then I went shopping for turkey day, a really nice guy helped me with my bags. Had lunch with my husband today – he also helped me with the bags. (Grant JE, 11/26/02)

At the end of the day, our careers as STEM professionals are one element of our total life. There are often a range of emotions and interactions that are managed in the course of a single day. When the interactions are positive, it feels good.

Yet another concept in interpersonal interactions, the sense of belonging, is described in Ong's work: "*Overwhelmingly, the research findings demonstrate that women of color, among other underrepresented groups, do not persist in STEM at the same rates as their White male counterparts due to social or interpersonal factors; in other words, women of color struggle and leave because they do not experience a sense of social belonging*" (Ong et al., 2018)

The following reflection by the author embodies the importance of belonging:

My colleagues made me laugh – they made me feel like I was one of them. (Grant JE, 3/24/06)

This is one element that is also difficult to manage/control, since we cannot "control" the behavior of others. The key here is to understand the impact that negative interactions can have on your ability to achieve the success that you seek. In some instances, the solution is to actually leave. Yes, leave the group or the organization to maintain your sanity, dignity and wellness. Before one takes that drastic step, a time of reflection can clarify the required actions.

Reflection, Rejuvenation, Regeneration and Revitalization (The 4 R's)

This set of terms came about from the author's extensive review of her journals; there were themes that emerged as she rediscovered painful elements of her journey that she had buried. These four R's are an essential component of the STEM journey; this entire chapter was borne out of an exercise dedicated to these very concepts. The four R's can be done solo, with trusted friends, family or in a coordinated group environment (e.g., group coaching). These four R's are ..*Almost like with stopping points along the way to examine your own life in the context of the unfolding drama!! (Grant JE, 8/10/05)*

The exercise of reflecting on the 4R's will be different for each of the readers of this chapter. In some instances- *you need to search for silence; take the parts of your life that need to be quieted now – not always out-front yelling and searching for answers. (Grant JE, 11/7/05)*

While we may not always be able to take long periods of respite to reflect on our STEM career, the incorporation of fun, soul-connecting elements can empower you along the way. Often, we can find them in unexpected places at seemingly inconvenient times. It's crucial to be open to when and where you can work on the four R's. The following reflects the positive elements in the author's STEM journey where she combined business and fun on the same trip to DC:

I attended a powerful Black Women in the Academy conference at Howard University in the Summer of 1999. I fellowshipped with women from STEM and non-STEM backgrounds. It didn't matter what our disciplines were, as we discussed our experiences, we found common experiential threads that wove through the fabric of our conversations. After the conference, I decided to work on personal and professional development while in the DC area: What a total blessing the meeting was! I also went to NSF to respond to a phone call about our XX proposal (it was eventually funded). I met with [my program officer] and it went well. I visited Auntie

Chris and that was a blessing – we talked, tried on clothes (she sells upscale fashion items), ate and had lots of fun (Grant JE, 6/28/99)

Balance, Transitions and Shifting Needs

Work-life balance is a misnomer, it really is work-life integration. In many situations it is important to understand when change is needed and if something's missing. We should also pay attention to shifting needs: ours and others. The author discovered that she spent a great deal of energy managing the various elements of work and home and writes:

Needs, wants and desires – how to separate them – what is really necessary, optional and just plain extra? How to discern what parts are really important to you, and to others? What is the central question that we seek to answer? (Grant JE, 2/16/04)

I don't want things to be perfect – just manageable so that I can feel like I'm on the right track with respect to my family. (Grant JE, 2/16/04)

Deep in my spirit I want to pick my son up each day this week at 3 p.m. Which means leaving the office at 2:40 p.m. at the latest to take a slow ride to his school and wait in the carpool lane. (Grant JE, 1/5/03)

In an era of continuous electronic connectivity, where emails, attachments and group online working platforms cause you to be engaged in a non-stop fashion with your career, we need to recognize both small and large transitions. In an industrial environment, a project endpoint could signal a transition to a new phase in your STEM career. For academics, the lines of demarcation occur between terms and over the summer. For example, in the summer faculty may not have regularly scheduled classes or meetings, if there are children or family responsibilities, there is an opportunity for increased flexibility.

Today is the last day of summer for this Mommy. Classes commence for me at 8:05 a.m. Thursday (Grant JE, 8/17/05)

At peace – end of the semester is near and I am looking forward to the church women's retreat on Friday and Saturday – it should really be a blessing. (Grant JE, 12/2/02)

Balance, transitions and shifting needs are some of the more difficult elements to manage in one's STEM Career Box, they are often highly personal and require the development of an individualized plan that accounts for things such as: family responsibilities, economics, assess to assistance, and a work environment that is conducive to change to meet the needs of its employees. Once again, one can glean a great deal of insight from networks of persons who are experiencing or are experts in these elements.

Shifting Definitions of Success

Our definitions of success may evolve as our careers evolve in the various realms of our lives (e.g., family, career, personal growth). For example, family dynamics can change based on different circumstances such as the birth of a child, the illness of a loved one, or the end of a relationship. How we successfully integrate these changes into our lives will be influenced by the stage of our STEM career. There can also be emerging levels of self-awareness relative to your interactions with other STEM professionals at different points in your career. While STEM excellence should always be a goal, the manner in which we interact with colleagues can and should evolve over time. The author found herself shifting as she progressed in her career:

Joy! I'm not sure why – yes, I am! I feel liberated from the excessive worry about what they think! They being my colleagues in the department. I am quite invisible to a number of my colleagues, Professor Y, Z, A etc. I've decided that I will not try to impress anyone anymore. (Grant JE, 1/6/03)

The shifting definition of success that are imposed by combinations external factors (e.g., job responsibility changes) and internal resolve will likely manifest on the STEM Success Cycle. It will be important to modify phases on the cycle to reflect one's changing definition of success.

Wellness, Well-being and Positive Psychology

One's mindset and overall wellness has a significant influence in our well-being as we pursue careers in STEM. The relationship that we have with the STEM journey can be overwhelming at times, taking a toll on one's wellness. The author recalls:

P.S. I didn't sleep last night at all- at least I don't think I did – I was always watching the clock and thinking about all the things I do and all the things pressing, mainly at work. I wrote a whole page list of items/issues -both short and longer term for me to keep track of. I started to address a few today. Buying that printer will help a lot with my work! (Grant JE, 1/12/03)

As you are considering the elements in your STEM Career Box, identify the wellness components that are beneficial to your journey. This may include a healthy lifestyle, emotional well-being, spiritual strength, and mental agility. The author regularly promotes the importance of getting assistance if any one of these components is out of balance. This assistance could range from an emotional counselor to a personal fitness trainer. The assistance could serve as a relieve valve for stress, frustration and pent up hostility.

Stress, Frustration and Pent Up Hostility

When you are the first Women of Color in a unit, group or even an entire college, understanding the "rules of engagement" and navigating the potential "professional potholes" can result in a degree of stress, frustration and pent up hostility. Anger is a normal, healthy human emotion. According to Psychology Today (*Anger*, n.d.):

Anger is one of the basic human emotions, as elemental as happiness, sadness, anxiety, or disgust. These emotions are tied to basic survival and were honed over the course of human history. Anger is related to the "fight, flight, or freeze" response of the sympathetic nervous system; it prepares humans to fight. Of course, anger too easily or frequently mobilized can undermine relationships and it can be deleterious to bodies in the long term. Prolonged release of the stress hormones that accompany anger can destroy neurons in areas of the brain associated with judgment and short-term memory, and weaken the immune system.

Over the years, sometimes the author found that the only place to fully express anger was in her journals:

ANGER … towards A, B, job and other issues…What can I say here. The details don't really matter. The issues are not important. The people are not relevant. Just know that I'm angry! Situations I can't control – people that are improper – how do I know what to say? How to handle them – how to approach a solution? (Grant JE, 8/10/05)

Where do you express your anger? Earlier in the chapter it was mentioned your online presence (e.g. social media) is a place where one could go to explore a timeline of the progression of your thoughts and responses to your elements of your STEM journey. This is not, of course, the place to display the raw elements of your anger and frustration. It is suggested that you: connect with your career mentor, a trusted friend or a peer to put the issues in the proper context; identify healthy solutions and explore options that may require formal engagement of personnel in your work environment.

Bias, Micro (And Macro) Aggressions

It was not until the second half of her career that the author learned words to describe some of her experiences. Terms like unconscious bias, critical race theory (J. DeCuir-Gunby, 2009; J. T. DeCuir-Gunby et al., 2013), intersectionality (Crenshaw, 1991), microaggressions and imposter syndrome can enable you to understand that "you are not alone" in your experiences. In the paper, "Understanding Female STEM Faculty Experiences of Subtle Gender Bias from Microaggressions Perspective", the authors define microaggressions as nuanced forms of insulting, and disrespectful communications that may occur during everyday exchanges (Yang & Carroll, 2016). While there are many forms of microaggressions, there are two that apply to WOC in STEM: gendered and racial microaggressions. Interestingly, *"unintentional microaggressions are usually the manifestation of unconscious biases, which are learned, automatic stereotypes that have the ability to affect our behavior*(Bastian, n.d.)." While one can not control the behavior of others, understanding the source of certain behaviors important when dealing with this element of your STEM Career Box. The promising news is that once something has a name, it can be addressed.

Inequitable Allocation of Resources and Opportunities

There are times when the allocation of resources and opportunities is done in a fair and equitable manner. Sometimes it is hard to know why decisions are made about resource distribution. In some instances, the inequity is an honest mistake, in other situations there is a difference in expectations between you, management and your STEM colleagues. At times you may think about seemingly outrageous solutions to issues of inequity; some are actually worth pursuing. Perhaps you have had a situation similar to this one experienced by the author:

I say I want to quit my job... again... today. It turns out that the newest XXX proposal does not have my name anywhere in the technical part. My husband says it's a blessing in disguise to encourage me to do my own thing (finally). I'm just tired of "fighting" for the things that should be mine – I'm also tired of being the second fiddle on these large centers. I presented at a [large center proposal] meeting today... where my topic was the last *thing on the agenda and not seriously considered. I also wasn't that excited about it either. not anymore. (Maybe I'll write a letter to Oprah Winfrey about my project...) Help me to exhibit wisdom in doing this – to have good style etc. I'm going to send my resume also – who knows it may open another door. (Grant JE, 1/9/03)*

Many years later, there was a realization that while there was an important role played in the multi-investigator effort, it was not properly aligned with what the author really wanted to do. It is hard to be passionate about something, something that you could lead and do well in, and have it relegated to a secondary status in a larger initiative. That is when one starts to develop the framework for the "making it you" approach to project planning and development. Mentors and coaches can be the key to a proper understanding of the politics behind the inequities and an approach to get what you need.

Long Term Anger, Jealousy

While we have already discussed some components of anger; there is no doubt what extreme negative emotions can do to elements of your mental, physical and emotional well-being.

If you have something inside you that is cancerous and it gets into the lymph node system it can move around and infect other parts of your body. Similarly, a cancer in your life can affect the others parts of your life. So, if you have a situation at work that is consuming your thought process and it starts to be a big part of your life – it <u>will</u> *affect the other areas of your life – such as home, personal relationships and your health. (Grant JE, 8/26/06)*

A high level of sustained anger can act like a cancer in your physical, mental, emotional and spiritual well-being. Oftentimes it is the result of, or becomes the cause of conflict. In the book "Facilitating a Collegial Department in Higher Education: Strategies for Success", Robert Cipriano (Cipriano & Cipriano, 2011) discussed methods for managing conflict in the department. They listed the following ways to manage conflict: (i) communicate, (ii) establish ground rules for airing disagreements, (iii) listen, (iv) monitor and control your emotions, (v) set the stage, (vi) understand exactly with the dispute is about, (vii) always address the issue, (viii) do your homework, (ix) try to handle issues informally whenever possible. He further indicated the importance of establishing a clear direction as to where you want the discussion to go, this means beginning with the end in mind. In particular, *communicating with an angry person should not be a competition to see "who wins". Being a good and effective communicator, you shift the exchange from the highly emotional to the rational.* (Cipriano & Cipriano, 2011) Anger alone is not always destructive, it can cause one to pivot to a constructive solution. Mental toughness is an element that can have a role in properly "managing" anger.

Mental Toughness

Mental toughness is yet another concept that can empower career success. During the author's rediscovery of the elements in her toolbox, she concluded that there was a great deal of resilience along the way. While resilience is a process, mental toughness is considered to be a personality trait (Moore, n.d.). Peggy Pritchard's chapter entitled, "Mental Toughness" in the Success Strategies book challenges the reader to adopt the following strategies (P. Pritchard, 2015): (i) mental strength, (ii) agility, (iii) flexibility and (iv) balance. Table 3 presents a more comprehensive summary of the strategies:

Table 3. Mental Toughness Strategies

Mental Strength	Agility	Flexibility	Balance
• Core strength • Strength of will	• Awareness • Stop escalation of tension • Let go and act mindfully	• Acceptance • Think creatively to solve problems and manage conflict	• Cultivate sound judgement • Maintain emotional equilibrium

While mental toughness is a powerful concept, it is clear that managing the elements of the STEM Career Box and the STEM Success Cycle require strategic attention to elements we can control. There is also a need for strategic allyship to change the culture and the climate for WOC in STEM.

It's Your Turn (Part 2: STEM Career Box)

Starting with the 14 items in Table 1, write down which elements you currently have in your STEM Career box. As stated earlier in the chapter, you may have other elements you want to include in the list. Some of these may even come from the Mental Toughness Strategies table. For each element, write a short description of how it manifests in your career box. If you are more advanced in your career like the author, consider reflecting on how these elements impacted and informed your career choices and how you can utilize that knowledge to move to the next phase of your career. Perhaps you also have a journal (or can start one) to use to reflect on the various elements in your past.

Now here is where the real work begins. Design and execute an approach to address and/or celebrate this element (you know it always helps to write things down). Table 4 represents a few examples:

Table 4. Sample elements, manifestation and executable approaches

Box Element	Identify How It Manifests and Question to Be Answered	Create and Execute Approach to Address/Celebrate
Wise Counsel from Coaches/Mentors	A number of my seasoned coaches and mentors are retiring or have passed away. *How do I celebrate their legacy, grow in their wisdom and fill the void left by their absence?*	• Find the salient points from your journals that captured their words of wisdom sowed into your life. • Set up times to record a few "career reflecting" conversations with them. • Figure out ways to incorporate their wisdom and legacy into my own wise counsel to others.
Unfinished Business	There is a next in my career which I need to strategically plan for and coordinate with the other components of my life. *What is the best way to accomplish this? What is the unfinished business in my career?*	• Take time to reflect and explore your early writings, listen to the "winds of STEM" to see where the gaps are in the profession, and see where you can make the greatest impact (and have fun!). • Wisdom from Journal Entry *Prepare for promotion – don't just seek out positions. Get ready for a new set of responsibilities. (Grant JE, 1/2/06)*
Unhealthy competition	STEM is a hyper-competitive field and at times the competition is even fiercer at the "top" in the leadership realm. *How can one ensure that the pursuit of opportunities or the incorporation of different projects into a STEM career portfolio is properly motivated?*	• Go back to the beginning, identify your ultimate career goals, look at the progression in your CV, and list your passions in the STEM realm. • Explore the career paths of mentors, and role models to find the elements you want to incorporate into your career portfolio. • Pay attention to the diversity of STEM positions in a variety of segments (academia, industry, government, etc.) where mutually beneficial opportunities may reside.

Finally, look at the STEM Success Cycle and identify how the elements of your STEM Career Box can be utilized in a creative manner to work on your STEM Success Cycle.

CONCLUSION

There are often elements at the beginning (and in the later stages) of our careers that are wonderful and rewarding. There are also elements that can be challenging in ways that help us to grow intellectually and mentally. The degree of control that we have to modify situations, behaviors, culture and policies is limited early in our career. Your own resilience and eventual ability to sit at the STEM table as an influencer leading change is an achievable goal. In a final reflection, the author reflects on her own STEM Success Cycle in the context of a 30-year STEM career through the lens of a journal entry from 15 years ago while riding on the DC Metro:

I made it – through the academic system in spite of the hostile stuff that was done to me. My career train moved through the ice, storms, mudslides and hills and valleys. Sometimes, the train had to stop, I thought we were stuck or stalled. But actually, we were stopped so the conductor could instruct the "workers" to get stuff off of the track. This enabled me to move forward again. Sometimes it looked like someone else's train was moving faster than mine and that their tracks were clear. But I didn't realize that they had a different conductor and were headed to different destinations. (Grant JE, 7/14/05)

The positive and the negative elements of the author's own STEM Career journey (while learning the rules and paying her dues) can all be traced back to elements of the literature on elements of challenge and success for WOC in STEM. The elements have prepared her to execute her current position as the Associate Dean of Faculty Advancement in the College of Engineering. Empowering faculty requires the ability to convert negative energy to positive outcomes while building on and celebrating the overwhelmingly positive elements in the STEM career boxes of the faculty. In hindsight, she believes that the positive and negative elements in her box were a combination of elements picked up, elements people threw into her box and even some that were already in the box.

Ultimately – you must recognize who you are and what you want to become. It's critically important to evaluate your surroundings and know what you are up against – so you can adjust and create your own realism for success. This can be done in the context of the extensive literature on WOC in STEM in addition to the summary recommendations from studies, conferences and thinktanks on the same topic. At the end of your life – there will be an obituary that summarizes your accomplishments and the impact of your life on others and on the world. Keeping that in mind is important – but it does not have to drive your daily actions.

In reflection, the author does not remember all the details of her own STEM life experiences. She knows that there were hard times, fun times and what could be considered interesting times in the life of a STEM professional. The ways in which the personal elements intersected with the professional elements in the authors career changed over time. She is still working on her STEM Career Box, discarding and adding elements as needed; she is also embarking on new phases in her STEM Success Cycle. The learned perspectives gained from experiences and interactions with others coupled with the infrastructure that manages our careers provides a core for decision-making. You can now utilize your STEM Career Box to ensure a positive outcome in your STEM Success Cycles.

ACKNOWLEDGMENT

This research received no specific grant from any funding agency in the public, commercial, or not-for-profit sectors.

The author is grateful for the input and insights from Barbara Smith and Julie Ivy. She also acknowledges the scholarly conversations over several years by her NSF ADVANCE co-PIs.

REFERENCES

ACS Women Chemists of Color Program. (n.d.). Retrieved January 31, 2019, from https://www.acs.org/content/acs/en/membership-and-networks/acs/welcoming/diversity/women-chemists-of-color.html

Anger. (n.d.). *Psychology Today*. Retrieved January 31, 2020, from https://www.psychologytoday.com/basics/anger

Camacho, M. M., & Lord, S. M. (2011). "Microaggressions" in engineering education: Climate for Asian, Latina and White women. *2011 Frontiers in Education Conference (FIE)*, S3H-1. 10.1109/FIE.2011.6142970

Carrigan, C., Kwak Tanguay, S., Yen, J., Ivy, J. S., Margherio, C., Riskin, E., ... Horner-Devine, C. (2018). *Building and Breaching Boundaries: An Intersectional Coherent Group Approach to Advancing Women Faculty in Engineering. 2018 ASEE Annual Conference & Exposition*, Salt Lake City, UT.

Cipriano, R. E., & Cipriano, R. E. (2011). *Facilitating a Collegial Department in Higher Education: Strategies for Success*. John Wiley & Sons, Incorporated. https://ebookcentral.proquest.com/lib/ncsu/detail.action?docID=697647

Crenshaw, K. (1991). Demarginalizing the Intersection of Race and Sex: A Black Feminist Critique of Antidiscrimination Doctrine, Feminist Theory, and Antiracist Politics [1989]. In Feminist Legal Theory (1st ed., Vol. 1–Book, Section, pp. 57–80). Routledge. doi:10.4324/9780429500480-5

DeCuir-Gunby, J. (2009). The Emotionality of Women Professors of Color in Engineering: A Critical Race Theory and Critical Race Feminism Perspective. In P. A. Schutz & M. Zembylas (Eds.), *Advances in Teacher Emotion Research* (pp. 323–342). Springer, US; doi:10.1007/978-1-4419-0564-2_16

DeCuir-Gunby, J. T., Grant, C., & Gregory, B. B. (2013). Exploring Career Trajectories for Women of Color in Engineering: The Experiences of African American and Latina Engineering Professors. *Journal of Women and Minorities in Science and Engineering*, *19*(3), 209–225. doi:10.1615/JWomenMinorScienEng.2013005769

Definition of SUCCESS2. (n.d.). Retrieved January 31, 2020, from https://www.lexico.com/en/definition/success

Definition of SUCCESS. (n.d.). Retrieved January 31, 2020, from https://www.merriam-webster.com/dictionary/success

Definition of SUCCESSFUL. (n.d.). Retrieved January 31, 2020, from https://www.merriam-webster.com/dictionary/successful

Fletcher, T., Ross, M., Tolbert, D., Holly, J., Cardella, M., Godwin, A., & DeBoer, J. (2017). *Ignored Potential: A Collaborative Roadmap for Increasing African American Women in Engineering*. National Society of Black Engineers. https://www.nsbe.org/getattachment/News-Media/NSBE-News/ignored-potential/NSBE-Ignored-Potential-Whitepaper-2-27-17.PDF.aspx

Goleman, D. (2014). What Makes a Leader. *Harvard Business Review*.

Grant, C. S. (2019). Mentors, Mentors Everywhere: Weaving Informal and Formal Mentoring into a Robust Chemical Sciences Mentoring Quilt. In Growing Diverse STEM Communities: Methodology, Impact, and Evidence (Vol. 1328, pp. 229–254). American Chemical Society. doi:10.1021/bk-2019-1328.ch015

Hill, C., Corbett, C., St. Rose, A., & American Association of University Women. (2010). *Why So Few? Women in Science, Technology, Engineering, and Mathematics*. American Association of University Women. https://www.aauw.org/research/why-so-few/

Horner-Devine, C. (2019). Community and Career Development: LATTICE for Early-career Underrepresented Minority Women. *POSTDOCket, 16*(10). https://www.nationalpostdoc.org/page/POSTDOCket_1610#

Johnson, J. G. (2015). Negotiating with Emotional Intelligence. In P. A. Pritchard & C. S. Grant (Eds.), *Success Strategies From Women in STEM* (2nd ed., pp. 299–333). Academic Press.

Lent, J. (2009). Journaling Enters the 21st Century: The Use of Therapeutic Blogs in Counseling. *Journal of Creativity in Mental Health, 4*(1), 67–73. doi:10.1080/15401380802705391

Malcom, L. E., & Malcom, S. M. (2011). The Double Bind: The Next Generation. *Harvard Educational Review, 81*(2), 389. doi:10.17763/haer.81.2.a84201x508406327

Malcom, S. M., Hall, P. Q., & Brown, J. W. (1976). *The Double Bind: The Price of Being a Minority Woman in Science*. http://web.mit.edu/cortiz/www/Diversity/1975-DoubleBind.pdf

McCord, C., Rigg, L., Levin, A., Coller, B., & Reynolds, J. (n.d.). *Navigate, Balance, and Retain: Developing Success in the Mid-Career for Female STEM Faculty (NFS 1015932)*. Academic Press.

Moore. (n.d.). *Resilience Training: How to Master Mental Toughness and Thrive*. Retrieved March 27, 2020, from https://positivepsychology.com/resilience-training/

Nave, F. M., & Williams, R. I. (2018). *Sister Outsider: Advancing Women Faculty in Engineering and Technology at Historically Black Colleges and Universities*. ACS Annual Meeting, Boston, MA.

Obiomon, P. H., Tickles, V. C., Wowo, A. H., & Holland-Hunt, S. (2007). *Advancement of Women of Color in Science, Technology, Engineering, and Math (STEM) Disciplines)*. Faculty Resource Network.

Ong, M., Smith, J. M., & Ko, L. T. (2018). Counterspaces for women of color in STEM higher education: Marginal and central spaces for persistence and success. *Journal of Research in Science Teaching, 55*(2), 206–245. doi:10.1002/tea.21417

Pritchard, P. (2015). Mental Toughness. In P. A. Pritchard & C. S. Grant (Eds.), *Microaggressions: Good Intent Is Not Good Enough*. Retrieved March 27, 2020, from https://www.forbes.com/sites/rebekahbastian/2019/06/10/microaggressions-good-intent-is-not-good-enough/#67aae5c87fb1

Steele, C. M., & Aronson, J. (1995). Stereotype threat and the intellectual test performance of African Americans. *Journal of Personality and Social Psychology, 69*(5), 797–811. doi:10.1037/0022-3514.69.5.797 PMID:7473032

Wilkins-Yel, K. G., Hyman, J., & Zounlome, N. O. O. (2019). Linking intersectional invisibility and hypervisibility to experiences of microaggressions among graduate women of color in STEM. *Journal of Vocational Behavior, 113*, 51–61. doi:10.1016/j.jvb.2018.10.018

Women Chemists Committee. (n.d.). Retrieved January 31, 2019, from https://www.womenchemists.sites.acs.org/

Yang, Y. L., & Carroll, D. W. (2016). Understanding Female STEM Faculty Experiences of Subtle Gender Bias from Microaggressions Perspective. *Association for Engineering Education - Engineering Library Division Papers*. doi:10.18260/p.27098

Related Readings

To continue IGI Global's long-standing tradition of advancing innovation through emerging research, please find below a compiled list of recommended IGI Global book chapters and journal articles in the areas of green cities, environmental management, and sustainable urban development. These related readings will provide additional information and guidance to further enrich your knowledge and assist you with your own research.

Acholonu, U., Dickinson, J., Amato, D., & Pinkard, N. (2017). Minecraft Server Project: Designing Synchronous Blended Learning Environments to Support Distributed Mentorship. In Y. Rankin & J. Thomas (Eds.), *Moving Students of Color from Consumers to Producers of Technology* (pp. 197–219). Hershey, PA: IGI Global. doi:10.4018/978-1-5225-2005-4.ch010

Adewusi, A. O. (2019). Skill Acquisition and Entrepreneurial Intention Among Corp Members in Southwestern Nigeria. In S. Bala & P. Singhal (Eds.), *Gender Issues in Technical and Vocational Education Programs* (pp. 183–199). Hershey, PA: IGI Global. doi:10.4018/978-1-5225-8443-8.ch009

Adjin, D., & Muat, H. (2019). The Prospect of Post-Adoption Satisfaction and the Digital Gender Divide. In I. Williams, O. Millward, & R. Layton (Eds.), *Gender Gaps and the Social Inclusion Movement in ICT* (pp. 192–211). Hershey, PA: IGI Global. doi:10.4018/978-1-5225-7068-4.ch010

Aggrawal, A. (2019). Work-Family Balance From Women Technopreneurs' Perspectives: A Qualitative Enquiry. In F. Tomos, N. Kumar, N. Clifton, & D. Hyams-Ssekasi (Eds.), *Women Entrepreneurs and Strategic Decision Making in the Global Economy* (pp. 195–208). Hershey, PA: IGI Global. doi:10.4018/978-1-5225-7479-8.ch010

Akanbi, L. B. (2019). Undermining Leadership Effectiveness. In U. Thomas (Ed.), *Navigating Micro-Aggressions Toward Women in Higher Education* (pp. 144–166). Hershey, PA: IGI Global. doi:10.4018/978-1-5225-5942-9.ch007

Aktepe, S. O. (2020). Gender Inequality in Work Organizations: What HRM Practices Mean for Gender Inequality. In B. Tan (Ed.), *Macro and Micro-Level Issues Surrounding Women in the Workforce: Emerging Research and Opportunities* (pp. 1–36). Hershey, PA: IGI Global. doi:10.4018/978-1-5225-9163-4.ch001

Al-Rashed, A., & Al-Gilani, M. (2020). Gender Equity in Surgical Academia in Kuwait and the Arabian Gulf Region. In M. Bellini & V. Papalois (Eds.), *Gender Equity in the Medical Profession* (pp. 237–251). Hershey, PA: IGI Global. doi:10.4018/978-1-5225-9599-1.ch017

Al-wazedi, U. (2020). Breaking Silence Through Gender Jihad: Muslim Women and the #MeToo Movement. In B. Glimps & T. Ford (Eds.), *MeToo Issues in Religious-Based Institutions and Organizations* (pp. 225–246). Hershey, PA: IGI Global. doi:10.4018/978-1-5225-9195-5.ch009

Alves, H. (2020). Gender Equity in the Medical Profession as a Democratic Culture: The Portuguese Experience. In M. Bellini & V. Papalois (Eds.), *Gender Equity in the Medical Profession* (pp. 199–212). Hershey, PA: IGI Global. doi:10.4018/978-1-5225-9599-1.ch014

Aranda, A. S., & Jerrard, M. A. (2019). A Comparison Between Australia and Chile of Factors Facing Women Engineers and ICT Professionals in Their Careers. In I. Williams, O. Millward, & R. Layton (Eds.), *Gender Gaps and the Social Inclusion Movement in ICT* (pp. 1–23). Hershey, PA: IGI Global. doi:10.4018/978-1-5225-7068-4.ch001

Aybas, M., & Özçelik, G. (2020). Entrepreneurship as a Career Choice for Women: A Successful Case Study From Turkey. In B. Tan (Ed.), *Macro and Micro-Level Issues Surrounding Women in the Workforce: Emerging Research and Opportunities* (pp. 89–108). Hershey, PA: IGI Global. doi:10.4018/978-1-5225-9163-4.ch004

Bala, S., & Singhal, P. (2019). Gender-Responsive TVET Framework: An Indian Perspective. In S. Bala & P. Singhal (Eds.), *Gender Issues in Technical and Vocational Education Programs* (pp. 235–258). Hershey, PA: IGI Global. doi:10.4018/978-1-5225-8443-8.ch012

Bernhardt, S., Braun, P., & Thomason, J. (2019). SocialIT. In *Gender Inequality and the Potential for Change in Technology Fields* (pp. 258–289). Hershey, PA: IGI Global. doi:10.4018/978-1-5225-7975-5.ch008

Bernhardt, S., Braun, P., & Thomason, J. (2019). Differences. In *Gender Inequality and the Potential for Change in Technology Fields* (pp. 290–327). Hershey, PA: IGI Global. doi:10.4018/978-1-5225-7975-5.ch009

Bobb, K., & Brown, Q. (2017). Access, Power, and the Framework of a CS Education Ecosystem. In Y. Rankin & J. Thomas (Eds.), *Moving Students of Color from Consumers to Producers of Technology* (pp. 245–260). Hershey, PA: IGI Global. doi:10.4018/978-1-5225-2005-4.ch012

Bradshaw, R. D. (2020). Queering the Marianist Charism: Narratives Offer Insights for Change. In B. Glimps & T. Ford (Eds.), *MeToo Issues in Religious-Based Institutions and Organizations* (pp. 76–105). Hershey, PA: IGI Global. doi:10.4018/978-1-5225-9195-5.ch004

Bragg, C. B. (2020). Gender Relations in the Black Church: Pentecostal Ecclesiology and Women's Leadership Roles in Transition. In B. Glimps & T. Ford (Eds.), *MeToo Issues in Religious-Based Institutions and Organizations* (pp. 106–142). Hershey, PA: IGI Global. doi:10.4018/978-1-5225-9195-5.ch005

Brown, K. H. (2019). From PWI to HBCU: When the Oppressed Takes on the Characteristics of the Oppressor. In U. Thomas (Ed.), *Navigating Micro-Aggressions Toward Women in Higher Education* (pp. 53–72). Hershey, PA: IGI Global. doi:10.4018/978-1-5225-5942-9.ch003

Carlin, P. M. (2020). Parental Leave During Surgical Training: Gender Equity, Parental Leave, Parental Policies, Childbearing, Conciliation, Surgery. In M. Bellini & V. Papalois (Eds.), *Gender Equity in the Medical Profession* (pp. 16–28). Hershey, PA: IGI Global. doi:10.4018/978-1-5225-9599-1.ch002

Carrim, N. M., & Senne, Y. (2019). Life Context Model, Intersectionality, and Black Feminist Epistemology: Women Managers in Africa. In E. Pereira & P. Paoloni (Eds.), *Handbook of Research on Women in Management and the Global Labor Market* (pp. 136–155). Hershey, PA: IGI Global. doi:10.4018/978-1-5225-9171-9.ch007

Carvalho, I., Costa, C., & Torres, A. (2019). Gender Awareness and Women Managers in Tourism: Perceptions of Inequality and What Could Be Done. In E. Pereira & P. Paoloni (Eds.), *Handbook of Research on Women in Management and the Global Labor Market* (pp. 218–238). Hershey, PA: IGI Global. doi:10.4018/978-1-5225-9171-9.ch011

Chandra, A. (2019). Gender Gap in Skill Development: An Analytical Review. In S. Bala & P. Singhal (Eds.), *Gender Issues in Technical and Vocational Education Programs* (pp. 1–22). Hershey, PA: IGI Global. doi:10.4018/978-1-5225-8443-8.ch001

Chandra, L., Downey, C. L., Svavarsdottir, H. S., Skinner, H., & Young, A. L. (2020). Gender Equity in Medical Leadership. In M. Bellini & V. Papalois (Eds.), *Gender Equity in the Medical Profession* (pp. 94–109). Hershey, PA: IGI Global. doi:10.4018/978-1-5225-9599-1.ch007

Cheng, S. (2020). Women in Leadership From the Perspective of a Chinese-American Professional. In B. Tan (Ed.), *Macro and Micro-Level Issues Surrounding Women in the Workforce: Emerging Research and Opportunities* (pp. 190–209). Hershey, PA: IGI Global. doi:10.4018/978-1-5225-9163-4.ch008

Cinar, K. (2019). Women's Entrepreneurship in Patriarchal Societies: The Case of Women's Cooperatives in Turkey. In F. Tomos, N. Kumar, N. Clifton, & D. Hyams-Ssekasi (Eds.), *Women Entrepreneurs and Strategic Decision Making in the Global Economy* (pp. 79–98). Hershey, PA: IGI Global. doi:10.4018/978-1-5225-7479-8.ch005

Deka, S. J. (2019). Vocational Education and Training for Women: A Gender Divide Perspective. In S. Bala & P. Singhal (Eds.), *Gender Issues in Technical and Vocational Education Programs* (pp. 138–157). Hershey, PA: IGI Global. doi:10.4018/978-1-5225-8443-8.ch007

Denner, J., Martinez, J., & Thiry, H. (2017). Strategies for Engaging Hispanic/Latino Youth in the US in Computer Science. In Y. Rankin & J. Thomas (Eds.), *Moving Students of Color from Consumers to Producers of Technology* (pp. 24–48). Hershey, PA: IGI Global. doi:10.4018/978-1-5225-2005-4.ch002

Divine, T. F. (2019). The Black One: Microaggressions in a Criminal Justice Program. In U. Thomas (Ed.), *Navigating Micro-Aggressions Toward Women in Higher Education* (pp. 167–180). Hershey, PA: IGI Global. doi:10.4018/978-1-5225-5942-9.ch008

Donahoo, S. (2019). Owning Black Hair: The Pursuit of Identity and Authenticity in Higher Education. In U. Thomas (Ed.), *Navigating Micro-Aggressions Toward Women in Higher Education* (pp. 73–95). Hershey, PA: IGI Global. doi:10.4018/978-1-5225-5942-9.ch004

Eglash, R., Babbitt, W., Bennett, A., Bennett, K., Callahan, B., Davis, J., ... Tully, K. (2017). Culturally Situated Design Tools: Generative Justice as a Foundation for STEM Diversity. In Y. Rankin & J. Thomas (Eds.), *Moving Students of Color from Consumers to Producers of Technology* (pp. 132–151). Hershey, PA: IGI Global. doi:10.4018/978-1-5225-2005-4.ch007

Erete, S., Martin, C. K., & Pinkard, N. (2017). Digital Youth Divas: A Program Model for Increasing Knowledge, Confidence, and Perceptions of Fit in STEM amongst Black and Brown Middle School Girls. In Y. Rankin & J. Thomas (Eds.), *Moving Students of Color from Consumers to Producers of Technology* (pp. 152–173). Hershey, PA: IGI Global. doi:10.4018/978-1-5225-2005-4.ch008

Faďoš, M., & Bohdalová, M. (2019). Evolution of the Employment Gender Inequality. In E. Pereira & P. Paoloni (Eds.), *Handbook of Research on Women in Management and the Global Labor Market* (pp. 21–45). Hershey, PA: IGI Global. doi:10.4018/978-1-5225-9171-9.ch002

Faulkner, P. E., Begum, S., & English, C. W. (2019). How Bangladesh Is Increasing the Employability of Females With Technical Training Programs. In S. Bala & P. Singhal (Eds.), *Gender Issues in Technical and Vocational Education Programs* (pp. 158–182). Hershey, PA: IGI Global. doi:10.4018/978-1-5225-8443-8.ch008

Ferrari, F. (2019). The Daughter's Career in Family Firms: A Reflection on the Cultural/Contextual Aspects. In E. Pereira & P. Paoloni (Eds.), *Handbook of Research on Women in Management and the Global Labor Market* (pp. 156–176). Hershey, PA: IGI Global. doi:10.4018/978-1-5225-9171-9.ch008

Fisherman, S. (2020). Body Image and Emotional Well Being Among Gay and Heterosexual Religious Young Men: Modern Orthodox and Ultra-Orthodox Jews. In B. Glimps & T. Ford (Eds.), *MeToo Issues in Religious-Based Institutions and Organizations* (pp. 195–224). Hershey, PA: IGI Global. doi:10.4018/978-1-5225-9195-5.ch008

Floris, M., Cinzia, D., & Angela, D. (2019). Women's Roles in Family Businesses: Specifics, Contradictions, and Reflections. In E. Pereira & P. Paoloni (Eds.), *Handbook of Research on Women in Management and the Global Labor Market* (pp. 177–199). Hershey, PA: IGI Global. doi:10.4018/978-1-5225-9171-9.ch009

Ford, T. N., & Glimps, B. J. (2020). The Sexual Abuse of African Nuns: No Sanctuary for African Women in the Sisterhood. In B. Glimps & T. Ford (Eds.), *MeToo Issues in Religious-Based Institutions and Organizations* (pp. 247–268). Hershey, PA: IGI Global. doi:10.4018/978-1-5225-9195-5.ch010

Galeshi, R. (2017). Engineering Education for All: Increasing Access to Engineering Education for Men and Women across the World through Distance Learning. *International Journal of Online Pedagogy and Course Design*, *7*(2), 35–47. doi:10.4018/IJOPCD.2017040103

Gans, E. A. (2020). Understanding the Value of Gender Equity in Healthcare and How to Pursue It. In M. Bellini & V. Papalois (Eds.), *Gender Equity in the Medical Profession* (pp. 64–74). Hershey, PA: IGI Global. doi:10.4018/978-1-5225-9599-1.ch005

Gardner-McCune, C., & Jimenez, Y. (2017). Historical App Developers: Integrating CS into K-12 through Cross-Disciplinary Projects. In Y. Rankin & J. Thomas (Eds.), *Moving Students of Color from Consumers to Producers of Technology* (pp. 85–112). Hershey, PA: IGI Global. doi:10.4018/978-1-5225-2005-4.ch005

George, T. O., Onwumah, A. C., Fagbohun, M. O., Adebayo, M. E., & Olonade, O. Y. (2019). Gender Differences in ICT Studies: A Study of Selected Public Secondary Schools in Ogun State, Nigeria. In I. Williams, O. Millward, & R. Layton (Eds.), *Gender Gaps and the Social Inclusion Movement in ICT* (pp. 147–169). Hershey, PA: IGI Global. doi:10.4018/978-1-5225-7068-4.ch008

Ginsberg, F., Davis, J., & Simms, A. (2019). Women in Higher Education Leadership: Challenges Are Many While Opportunities Are Few. In H. Schnackenberg & D. Simard (Eds.), *Challenges and Opportunities for Women in Higher Education Leadership* (pp. 219–237). Hershey, PA: IGI Global. doi:10.4018/978-1-5225-7056-1.ch013

Golic, Z. (2019). Problem of Financing Women Entrepreneurs: Experience of Women Entrepreneurs in Post-Conflict Bosnia and Herzegovina. In F. Tomos, N. Kumar, N. Clifton, & D. Hyams-Ssekasi (Eds.), *Women Entrepreneurs and Strategic Decision Making in the Global Economy* (pp. 278–304). Hershey, PA: IGI Global. doi:10.4018/978-1-5225-7479-8.ch014

Gonyea, L. T. (2019). Women in Higher Education Leadership: Exploring the Intersections of Race and Gender. In H. Schnackenberg & D. Simard (Eds.), *Challenges and Opportunities for Women in Higher Education Leadership* (pp. 44–60). Hershey, PA: IGI Global. doi:10.4018/978-1-5225-7056-1.ch004

Gorkey, S. (2019). Women's Employment in Turkey's ICT Sector: An Examination From a Social Inclusion Perspective. In I. Williams, O. Millward, & R. Layton (Eds.), *Gender Gaps and the Social Inclusion Movement in ICT* (pp. 63–86). Hershey, PA: IGI Global. doi:10.4018/978-1-5225-7068-4.ch004

Grant, Y. (2020). Sexism in Medical Communication. In M. Bellini & V. Papalois (Eds.), *Gender Equity in the Medical Profession* (pp. 75–93). Hershey, PA: IGI Global. doi:10.4018/978-1-5225-9599-1.ch006

Grossman, R. C. (2020). The Gender Pay Gap in Medicine: Causes and Solutions. In M. Bellini & V. Papalois (Eds.), *Gender Equity in the Medical Profession* (pp. 110–127). Hershey, PA: IGI Global. doi:10.4018/978-1-5225-9599-1.ch008

Grossman, R. C. (2020). The Imposter Phenomenon in the Medical Profession. In M. Bellini & V. Papalois (Eds.), *Gender Equity in the Medical Profession* (pp. 148–166). Hershey, PA: IGI Global. doi:10.4018/978-1-5225-9599-1.ch011

Hargrave, C. P. (2019). Analyzing University Exploitation of Diversity to Legitimize Hiring Discrimination: A Black Woman Professor's Narrative. In U. Thomas (Ed.), *Navigating Micro-Aggressions Toward Women in Higher Education* (pp. 234–267). Hershey, PA: IGI Global. doi:10.4018/978-1-5225-5942-9.ch012

Harries, R. L., Nicholas, R. S., & Mohan, H. M. (2020). Supporting Surgeons to Have Families: Fertility, Pregnancy, Parental Leave, and Return to Work. In M. Bellini & V. Papalois (Eds.), *Gender Equity in the Medical Profession* (pp. 1–15). Hershey, PA: IGI Global. doi:10.4018/978-1-5225-9599-1.ch001

Hayes, C., & Graham, Y. N. (2020). Prophylaxis in Action: #MeToo for Women of Medical and Surgical Disciplines. In M. Bellini & V. Papalois (Eds.), *Gender Equity in the Medical Profession* (pp. 270–279). Hershey, PA: IGI Global. doi:10.4018/978-1-5225-9599-1.ch019

Heisler, C. A., Stephens, E. H., Temkin, S. M., & Miller, P. (2020). Gender Inequity in the United States Surgical Workforce. In M. Bellini & V. Papalois (Eds.), *Gender Equity in the Medical Profession* (pp. 252–269). Hershey, PA: IGI Global. doi:10.4018/978-1-5225-9599-1.ch018

Hernanz, Á. M., Herrera de la Muela, M., & González-Estecha, M. (2020). Impostor Syndrome as a Risk Factor in the Development of Psychiatric Symptoms in Female Doctors. In M. Bellini & V. Papalois (Eds.), *Gender Equity in the Medical Profession* (pp. 137–147). Hershey, PA: IGI Global. doi:10.4018/978-1-5225-9599-1.ch010

Hill, L. H., Wheat, C. A., Mandishona, T. C., & Blake, A. E. (2019). Women's Leadership Aspirations and Career Paths in Higher Education: Influence of Personal Factors. In H. Schnackenberg & D. Simard (Eds.), *Challenges and Opportunities for Women in Higher Education Leadership* (pp. 176–198). Hershey, PA: IGI Global. doi:10.4018/978-1-5225-7056-1.ch011

Honorato-Cia, C., De Hert, S., De Robertis, E., & Grigoras, I. (2020). Who Will Take Care of Me?: The Future of Human Resources in Anaesthesiology, Critical Care, and Emergency Medicine in Europe. In M. Bellini & V. Papalois (Eds.), *Gender Equity in the Medical Profession* (pp. 167–181). Hershey, PA: IGI Global. doi:10.4018/978-1-5225-9599-1.ch012

Huderson, A., & Huderson, B. (2019). Societal Factors and Workplace Perceptions: Understanding Social Determinants of Professional STEM Achievement and Persistence for Black Women. In U. Thomas & J. Drake (Eds.), *Women's Influence on Inclusion, Equity, and Diversity in STEM Fields* (pp. 1–21). Hershey, PA: IGI Global. doi:10.4018/978-1-5225-8870-2.ch001

Husband, T., & Brown, C. (2018). All My Life I Had to Fight: The Disproportionate Impact of School Disciplinary Policies on Black Girls in K-12 Contexts. In M. Grant (Ed.), *Equity, Equality, and Reform in Contemporary Public Education* (pp. 53–68). Hershey, PA: IGI Global. doi:10.4018/978-1-5225-4960-4.ch003

Hyams-Ssekasi, D., Stefan, A., Agboma, F., & Kumar, N. (2019). Determinants of Women's Entrepreneurial Attitude Across European Cultures. In F. Tomos, N. Kumar, N. Clifton, & D. Hyams-Ssekasi (Eds.), *Women Entrepreneurs and Strategic Decision Making in the Global Economy* (pp. 63–78). Hershey, PA: IGI Global. doi:10.4018/978-1-5225-7479-8.ch004

Ifedi, R. (2017). African-Born Female Academics in the U.S.: Experiences of Inclusion, Exclusion, and Access - Building Careers on Marginalized Identities. *International Journal of Bias, Identity and Diversities in Education*, *2*(1), 1–12. doi:10.4018/IJBIDE.2017010101

Inouye, T. M., Robinson, J. A., & Joshi, A. M. (2020). Does a Federal Glass Ceiling Have Differential Effects on Female and Male Technology Entrepreneurs? In B. Tan (Ed.), *Macro and Micro-Level Issues Surrounding Women in the Workforce: Emerging Research and Opportunities* (pp. 109–131). Hershey, PA: IGI Global. doi:10.4018/978-1-5225-9163-4.ch005

Iroaganachi, M. A., Ilo, P. I., Ekwueme, L., & YoungHarry, I. (2019). Knowledge Sharing Practices, ICT, Information Literacy, and STEM Career Path Choices Among Girls at the Secondary School Level: Girls and STEM Education. In U. Thomas, & J. Drake (Eds.), *Women's Influence on Inclusion, Equity, and Diversity in STEM Fields* (pp. 227-252). Hershey, PA: IGI Global. doi:10.4018/978-1-5225-8870-2.ch009

Jackson, H. C. (2019). Overcoming the Layers of Obstacles: The Journey of a Female African American Physicist to Achieve Equity, Diversity, and Inclusiveness. In U. Thomas & J. Drake (Eds.), *Women's Influence on Inclusion, Equity, and Diversity in STEM Fields* (pp. 76–107). Hershey, PA: IGI Global. doi:10.4018/978-1-5225-8870-2.ch004

Jafaar, A. H., Juma, M. Y., Habib, J. M., & Al-Sartawi, A. M. (2019). Bahraini Women's Voices in the Boardroom: The Effect of Gender Diversity on Firm Performance. In E. Pereira & P. Paoloni (Eds.), *Handbook of Research on Women in Management and the Global Labor Market* (pp. 336–350). Hershey, PA: IGI Global. doi:10.4018/978-1-5225-9171-9.ch017

Johnson, N. N., & Johnson, T. L. (2019). Microaggressions: An Introduction. In U. Thomas (Ed.), *Navigating Micro-Aggressions Toward Women in Higher Education* (pp. 1–22). Hershey, PA: IGI Global. doi:10.4018/978-1-5225-5942-9.ch001

Jones, V. N., Dopwell, D. M., & Curry, L. C. (2020). Silenced, Shamed, and Scatted: Black Feminist Perspective on Sexual Trauma and Treatment With African American Female Survivors. In B. Glimps & T. Ford (Eds.), *MeToo Issues in Religious-Based Institutions and Organizations* (pp. 1–32). Hershey, PA: IGI Global. doi:10.4018/978-1-5225-9195-5.ch001

Jorge, A., Alves, H., Carvalho, F., Ferreira, M. A., & Machado, M. (2020). Gender Equity Evaluated by Five Successful Women in the Medical Profession. In M. Bellini & V. Papalois (Eds.), *Gender Equity in the Medical Profession* (pp. 182–198). Hershey, PA: IGI Global. doi:10.4018/978-1-5225-9599-1.ch013

Kalafatoğlu, Y. (2020). Women-Founded Start-Ups: Bigger, Better, Faster, More! In B. Tan (Ed.), *Macro and Micro-Level Issues Surrounding Women in the Workforce: Emerging Research and Opportunities* (pp. 58–88). Hershey, PA: IGI Global. doi:10.4018/978-1-5225-9163-4.ch003

Khamis, R., Barone, E., Sarea, A., & Hamdan, A. M. (2019). Board Gender Diversity and Firm Performance: Integrating Agency and Institutional Theory. In E. Pereira & P. Paoloni (Eds.), *Handbook of Research on Women in Management and the Global Labor Market* (pp. 302–311). Hershey, PA: IGI Global. doi:10.4018/978-1-5225-9171-9.ch015

Khamseh, Z. (2019). Power and Gender: A Comparative Cross-Cultural Study of Working Women. In E. Pereira & P. Paoloni (Eds.), *Handbook of Research on Women in Management and the Global Labor Market* (pp. 91–112). Hershey, PA: IGI Global. doi:10.4018/978-1-5225-9171-9.ch005

Kılıçaslan-Gökoğlu, S., & Öztürk, E. B. (2020). End of Occupational Segregation in the Nursing Profession: What Has Changed for Female Nurses? In B. Tan (Ed.), *Macro and Micro-Level Issues Surrounding Women in the Workforce: Emerging Research and Opportunities* (pp. 262–284). Hershey, PA: IGI Global. doi:10.4018/978-1-5225-9163-4.ch011

Kişi, N. (2019). Understanding Career Barriers of Women Executives: Glass Ceiling and Glass Cliff. In E. Pereira & P. Paoloni (Eds.), *Handbook of Research on Women in Management and the Global Labor Market* (pp. 200–217). Hershey, PA: IGI Global. doi:10.4018/978-1-5225-9171-9.ch010

Kose, T. (2019). Gender and Use of E-Government Services in Turkey: E-Government in Turkey. In I. Williams, O. Millward, & R. Layton (Eds.), *Gender Gaps and the Social Inclusion Movement in ICT* (pp. 130–146). Hershey, PA: IGI Global. doi:10.4018/978-1-5225-7068-4.ch007

Kurtz, R. S. (2019). Fortitude: A Study of African Americans in Surgery in New York City. In U. Thomas & J. Drake (Eds.), *Women's Influence on Inclusion, Equity, and Diversity in STEM Fields* (pp. 194–226). Hershey, PA: IGI Global. doi:10.4018/978-1-5225-8870-2.ch008

Layton, R. (2019). The Conundrum of Falling Participation of Women in Math and Computing Jobs: Observations From the USA and Denmark. In I. Williams, O. Millward, & R. Layton (Eds.), *Gender Gaps and the Social Inclusion Movement in ICT* (pp. 45–62). Hershey, PA: IGI Global. doi:10.4018/978-1-5225-7068-4.ch003

Lea, Y., & Butterfield, C. L. (2019). Praxis of the Teaching Profession: A Dialectic of Institutional Oppression and the Development of Pedagogy and Critical Consciousness. In U. Thomas (Ed.), *Navigating Micro-Aggressions Toward Women in Higher Education* (pp. 202–217). Hershey, PA: IGI Global. doi:10.4018/978-1-5225-5942-9.ch010

Leggett-Robinson, P. M., & Villa, B. C. (2019). Navigating the Landscape of the STEM Professoriate: Reflections and Insights From Women of Color. In U. Thomas & J. Drake (Eds.), *Women's Influence on Inclusion, Equity, and Diversity in STEM Fields* (pp. 22–65). Hershey, PA: IGI Global. doi:10.4018/978-1-5225-8870-2.ch002

Leung, L. Y. (2020). Representation of Gender Equality From the Perspective of the Medical Trainee and its Ripple Effect: Highlighting Gender Inequality in Medical Student Experiences. In M. Bellini & V. Papalois (Eds.), *Gender Equity in the Medical Profession* (pp. 29–50). Hershey, PA: IGI Global. doi:10.4018/978-1-5225-9599-1.ch003

Lewis, A. S. (2019). What's Respect Got to Do With It?: A Black Woman's Experience With the Role of Respect in Academia. In U. Thomas (Ed.), *Navigating Micro-Aggressions Toward Women in Higher Education* (pp. 181–201). Hershey, PA: IGI Global. doi:10.4018/978-1-5225-5942-9.ch009

Lituchy, T. R., Galperin, B. L., Melyoki, L., Senaji, T., & Punnett, B. J. (2020). African Women in Leadership: Lessons From the LEAD Research Project. In B. Tan (Ed.), *Macro and Micro-Level Issues Surrounding Women in the Workforce: Emerging Research and Opportunities* (pp. 161–189). Hershey, PA: IGI Global. doi:10.4018/978-1-5225-9163-4.ch007

Logan, S. R., & Dudley, H. S. (2019). The "Double-Whammy" of Being Black and a Woman in Higher Education Leadership. In H. Schnackenberg & D. Simard (Eds.), *Challenges and Opportunities for Women in Higher Education Leadership* (pp. 84–104). Hershey, PA: IGI Global. doi:10.4018/978-1-5225-7056-1.ch006

Lori, H., Hamdan, A. M., Sarea, A., & Al Shirawi, T. M. (2019). Women on Board and Firm Performance: Evidence From Kingdom of Bahrain. In E. Pereira & P. Paoloni (Eds.), *Handbook of Research on Women in Management and the Global Labor Market* (pp. 312–335). Hershey, PA: IGI Global. doi:10.4018/978-1-5225-9171-9.ch016

Lorraine-Lichtenstein, E. (2020). Gender Equity in Medicine in Sweden. In M. Bellini & V. Papalois (Eds.), *Gender Equity in the Medical Profession* (pp. 213–229). Hershey, PA: IGI Global. doi:10.4018/978-1-5225-9599-1.ch015

Loy, J. (2019). Women and Technology: Disrupting Leadership in Engineering Education. In H. Schnackenberg & D. Simard (Eds.), *Challenges and Opportunities for Women in Higher Education Leadership* (pp. 252–267). Hershey, PA: IGI Global. doi:10.4018/978-1-5225-7056-1.ch015

Luk, C. Y. (2019). A Human Rights-Based Approach to Bridge Gender Digital Divide: The Case Study of India. In I. Williams, O. Millward, & R. Layton (Eds.), *Gender Gaps and the Social Inclusion Movement in ICT* (pp. 24–44). Hershey, PA: IGI Global. doi:10.4018/978-1-5225-7068-4.ch002

Madaleno, M., Vieira, E. S., & Teodósio, J. P. (2019). Corporate Female Leadership Effects Considering PSI20 and IBEX35 Companies' Performance. In E. Pereira & P. Paoloni (Eds.), *Handbook of Research on Women in Management and the Global Labor Market* (pp. 280–301). Hershey, PA: IGI Global. doi:10.4018/978-1-5225-9171-9.ch014

Mallawaarachchi, C. (2019). Significant Influences of Digital Literacy to Promote Gender Equity and Equality in Education: Digital Literacy Perspective. In S. Bala & P. Singhal (Eds.), *Gender Issues in Technical and Vocational Education Programs* (pp. 218–234). Hershey, PA: IGI Global. doi:10.4018/978-1-5225-8443-8.ch011

Mampane, J. N. (2019). Youth Participation in the National Development Plan Through Technical and Vocational Education and Training: The Case of Young Black Women in South Africa. In S. Bala & P. Singhal (Eds.), *Gender Issues in Technical and Vocational Education Programs* (pp. 23–42). Hershey, PA: IGI Global. doi:10.4018/978-1-5225-8443-8.ch002

Margolis, J., Goode, J., & Flapan, J. (2017). A Critical Crossroads for Computer Science for All: "Identifying Talent" or "Building Talent," and What Difference Does It Make? In Y. Rankin & J. Thomas (Eds.), *Moving Students of Color from Consumers to Producers of Technology* (pp. 1–23). Hershey, PA: IGI Global. doi:10.4018/978-1-5225-2005-4.ch001

Martin, C. K., Pinkard, N., Erete, S., & Sandherr, J. (2017). Connections at the Family Level: Supporting Parents and Caring Adults to Engage Youth in Learning about Computers and Technology. In Y. Rankin & J. Thomas (Eds.), *Moving Students of Color from Consumers to Producers of Technology* (pp. 220–244). Hershey, PA: IGI Global. doi:10.4018/978-1-5225-2005-4.ch011

Mayakis, C. G., & Robinson, J. (2018). Girls in STEM K-12 Subjects: Exploring the Confidence and Hindrance of Pursuing STEM Careers. In M. Grant (Ed.), *Equity, Equality, and Reform in Contemporary Public Education* (pp. 110–125). Hershey, PA: IGI Global. doi:10.4018/978-1-5225-4960-4.ch006

McKinnon-Crowley, S. (2019). Women Community College Faculty Members: Current and Future Pathways to Leadership. In H. Schnackenberg & D. Simard (Eds.), *Challenges and Opportunities for Women in Higher Education Leadership* (pp. 238–251). Hershey, PA: IGI Global. doi:10.4018/978-1-5225-7056-1.ch014

McNally, S. A. (2020). Parallels Between Improving the Culture for Women in Surgery and a Surgical Approach to Improving Health for All. In M. Bellini & V. Papalois (Eds.), *Gender Equity in the Medical Profession* (pp. 128–136). Hershey, PA: IGI Global. doi:10.4018/978-1-5225-9599-1.ch009

Mitchell, R. L., Carr, E. A., Garcia, L. R., Steinman, L. E., Kowalski-Braun, M., & Beach, A. L. (2019). Holding Out for a Hero: Leadership Discourse in College and University Presidential Job Postings in a 20-Year Span. In H. Schnackenberg & D. Simard (Eds.), *Challenges and Opportunities for Women in Higher Education Leadership* (pp. 156–175). Hershey, PA: IGI Global. doi:10.4018/978-1-5225-7056-1.ch010

Moreland, M., & Thompson, T. (2019). Mentoring Female Leaders at Multiple Levels in One Higher Education Institution. In H. Schnackenberg & D. Simard (Eds.), *Challenges and Opportunities for Women in Higher Education Leadership* (pp. 20–28). Hershey, PA: IGI Global. doi:10.4018/978-1-5225-7056-1.ch002

Nacu, D. C., Martin, C. K., Sandherr, J., & Pinkard, N. (2017). An Interface Design Approach to Encouraging Online Contributions among Underrepresented Youth. In Y. Rankin & J. Thomas (Eds.), *Moving Students of Color from Consumers to Producers of Technology* (pp. 174–196). Hershey, PA: IGI Global. doi:10.4018/978-1-5225-2005-4.ch009

Ncube, R., & Chirisa, I. (2019). Female Entrepreneurship in Africa: An Inquiry Into the Influence of Rotating Savings and Credit Associations on Business Growth. In E. Pereira & P. Paoloni (Eds.), *Handbook of Research on Women in Management and the Global Labor Market* (pp. 259–279). Hershey, PA: IGI Global. doi:10.4018/978-1-5225-9171-9.ch013

Norwood, R. S. (2019). Molding Me in Their Image. In U. Thomas (Ed.), *Navigating Micro-Aggressions Toward Women in Higher Education* (pp. 218–233). Hershey, PA: IGI Global. doi:10.4018/978-1-5225-5942-9.ch011

Okezie, C. E., Alhamisi, J., & Glimps, B. J. (2020). The Promise for African American Male Students in Graduate Studies and Professional Development at Marygrove College. In B. Glimps & T. Ford (Eds.), *MeToo Issues in Religious-Based Institutions and Organizations* (pp. 33–55). Hershey, PA: IGI Global. doi:10.4018/978-1-5225-9195-5.ch002

Omeihe, K. O., Dakhan, S. A., Khan, M. S., Gustafsson, V., & Amoako, I. O. (2019). Riding the Wave: Understanding the Context of Female Entrepreneurship in Pakistan. In F. Tomos, N. Kumar, N. Clifton, & D. Hyams-Ssekasi (Eds.), *Women Entrepreneurs and Strategic Decision Making in the Global Economy* (pp. 20–39). Hershey, PA: IGI Global. doi:10.4018/978-1-5225-7479-8.ch002

Özçelik, G. (2020). Women's Participation in the Labor Market: A Comparative Analysis Between Turkey and the Europe and Central Asia Region. In B. Tan (Ed.), *Macro and Micro-Level Issues Surrounding Women in the Workforce: Emerging Research and Opportunities* (pp. 132–160). Hershey, PA: IGI Global. doi:10.4018/978-1-5225-9163-4.ch006

Özçelik, T. Ö. (2019). Sustainable Development With the Digitalization of Women: Economic Empowerment, Information Technology, and Women. In I. Williams, O. Millward, & R. Layton (Eds.), *Gender Gaps and the Social Inclusion Movement in ICT* (pp. 170–191). Hershey, PA: IGI Global. doi:10.4018/978-1-5225-7068-4.ch009

Patil, P. A., & Deshpande, Y. M. (2019). Women Entrepreneurship: A Journey Begins. In F. Tomos, N. Kumar, N. Clifton, & D. Hyams-Ssekasi (Eds.), *Women Entrepreneurs and Strategic Decision Making in the Global Economy* (pp. 99–118). Hershey, PA: IGI Global. doi:10.4018/978-1-5225-7479-8.ch006

Peer, K. (2019). Skill Challenge in Bengaluru's Garment Industry: A Gender Perspective. In S. Bala & P. Singhal (Eds.), *Gender Issues in Technical and Vocational Education Programs* (pp. 95–117). Hershey, PA: IGI Global. doi:10.4018/978-1-5225-8443-8.ch005

Pereira, E. T., & Salaris, S. (2019). The Evolution of the Role of Women in Labor Markets in Developed Economies. In E. Pereira & P. Paoloni (Eds.), *Handbook of Research on Women in Management and the Global Labor Market* (pp. 1–20). Hershey, PA: IGI Global. doi:10.4018/978-1-5225-9171-9.ch001

Pimpa, N., & Moore, T. (2019). Gender, CSR, and Mining: Perspectives From Thailand. In *Corporate Social Responsibility and the Inclusivity of Women in the Mining Industry: Emerging Research and Opportunities* (pp. 63–69). Hershey, PA: IGI Global. doi:10.4018/978-1-5225-3811-0.ch005

Pimpa, N., & Moore, T. (2019). Gender, CSR, and Mining: Perspectives From Lao PDR. In *Corporate Social Responsibility and the Inclusivity of Women in the Mining Industry: Emerging Research and Opportunities* (pp. 53–62). Hershey, PA: IGI Global. doi:10.4018/978-1-5225-3811-0.ch004

Pimpa, N., & Moore, T. (2019). Mining and Women: Business and Community. In *Corporate Social Responsibility and the Inclusivity of Women in the Mining Industry: Emerging Research and Opportunities* (pp. 70–78). Hershey, PA: IGI Global. doi:10.4018/978-1-5225-3811-0.ch006

Pradere, D., Ford, T. N., & Glimps, B. J. (2020). Beyond the Catholic Church: Child Sexual Abuse in Selected Other Religious Organizations. In B. Glimps & T. Ford (Eds.), *MeToo Issues in Religious-Based Institutions and Organizations* (pp. 170–194). Hershey, PA: IGI Global. doi:10.4018/978-1-5225-9195-5.ch007

Price-Williams, S., & Maätita, F. (2019). Critical Examination of Tokenism and Demands of Organizational Citizenship Behavior Among Faculty Women of Color. In U. Thomas (Ed.), *Navigating Micro-Aggressions Toward Women in Higher Education* (pp. 96–121). Hershey, PA: IGI Global. doi:10.4018/978-1-5225-5942-9.ch005

Puliatte, A. (2019). An Unlikely School Leader: Experiences of an Introverted Woman. In H. Schnackenberg & D. Simard (Eds.), *Challenges and Opportunities for Women in Higher Education Leadership* (pp. 144–155). Hershey, PA: IGI Global. doi:10.4018/978-1-5225-7056-1.ch009

Ramsay-Jordan, N. N. (2019). What Sports Do You Play? Actually, My Major Is Mathematics: Experiencing STEM as a Woman of African Descent. In U. Thomas & J. Drake (Eds.), *Women's Influence on Inclusion, Equity, and Diversity in STEM Fields* (pp. 66–75). Hershey, PA: IGI Global. doi:10.4018/978-1-5225-8870-2.ch003

Rankin, Y. A., & Thomas, J. O. (2017). Leveraging Food as the Context for Developing Computational Algorithmic Thinking in an Entry-Level College Course. In Y. Rankin & J. Thomas (Eds.), *Moving Students of Color from Consumers to Producers of Technology* (pp. 113–130). Hershey, PA: IGI Global. doi:10.4018/978-1-5225-2005-4.ch006

Roy, S. (2019). Challenges for Asian American and Immigrant Asian Women in the U.S. Higher Education Administration. In H. Schnackenberg & D. Simard (Eds.), *Challenges and Opportunities for Women in Higher Education Leadership* (pp. 105–124). Hershey, PA: IGI Global. doi:10.4018/978-1-5225-7056-1.ch007

Rudge, L. T. (2017). Different Experiences and Perceptions of Campus Climate Among Minority Students at a Predominantly White Institution. *International Journal of Bias, Identity and Diversities in Education*, *2*(1), 40–56. doi:10.4018/IJBIDE.2017010104

Russell, F. R. (2019). Successful Women Leaders: The Convergence of Personal and Professional Promotion. In H. Schnackenberg & D. Simard (Eds.), *Challenges and Opportunities for Women in Higher Education Leadership* (pp. 199–218). Hershey, PA: IGI Global. doi:10.4018/978-1-5225-7056-1.ch012

Sarmento, M., Dias, M. F., Amorim, M., Madaleno, M., & Pimentel, C. (2019). Transversal Competences Towards Employability in Female Students. In E. Pereira & P. Paoloni (Eds.), *Handbook of Research on Women in Management and the Global Labor Market* (pp. 63–90). Hershey, PA: IGI Global. doi:10.4018/978-1-5225-9171-9.ch004

Sawang, S., & Kivits, R. A. (2019). Born vs. Educated Entrepreneurs: Who Are Richer and Happier? In F. Tomos, N. Kumar, N. Clifton, & D. Hyams-Ssekasi (Eds.), *Women Entrepreneurs and Strategic Decision Making in the Global Economy* (pp. 258–276). Hershey, PA: IGI Global. doi:10.4018/978-1-5225-7479-8.ch013

Schnackenberg, H. L. (2019). motherscholar: MotherLeader. In H. Schnackenberg, & D. Simard (Eds.), Challenges and Opportunities for Women in Higher Education Leadership (pp. 29-43). Hershey, PA: IGI Global. doi:10.4018/978-1-5225-7056-1.ch003

Schnackenberg, H. L., & Simard, D. A. (2017). It's Not Easy, Being Queen. In *Challenges Facing Female Department Chairs in Contemporary Higher Education: Emerging Research and Opportunities* (pp. 23–34). Hershey, PA: IGI Global. doi:10.4018/978-1-5225-1891-4.ch003

Schnackenberg, H. L., & Simard, D. A. (2017). Moving Forward. In *Challenges Facing Female Department Chairs in Contemporary Higher Education: Emerging Research and Opportunities* (pp. 59–67). Hershey, PA: IGI Global. doi:10.4018/978-1-5225-1891-4.ch006

Schnackenberg, H. L., & Simard, D. A. (2017). Support Is Not Just for Pantyhose. In *Challenges Facing Female Department Chairs in Contemporary Higher Education: Emerging Research and Opportunities* (pp. 35–50). Hershey, PA: IGI Global. doi:10.4018/978-1-5225-1891-4.ch004

Scott, A., Martin, A., & McAlear, F. (2017). Enhancing Participation in Computer Science among Girls of Color: An Examination of a Preparatory AP Computer Science Intervention. In Y. Rankin & J. Thomas (Eds.), *Moving Students of Color from Consumers to Producers of Technology* (pp. 62–84). Hershey, PA: IGI Global. doi:10.4018/978-1-5225-2005-4.ch004

Scott, B. (2019). Afro-Caribbean Immigrants in STEM Careers: One Woman's Experience Teaching in a STEM Field. In U. Thomas & J. Drake (Eds.), *Women's Influence on Inclusion, Equity, and Diversity in STEM Fields* (pp. 143–168). Hershey, PA: IGI Global. doi:10.4018/978-1-5225-8870-2.ch006

Scuotto, V., Serravalle, F., Murray, A., & Viassone, M. (2019). The Shift Towards a Digital Business Model: A Strategic Decision for the Female Entrepreneur. In F. Tomos, N. Kumar, N. Clifton, & D. Hyams-Ssekasi (Eds.), *Women Entrepreneurs and Strategic Decision Making in the Global Economy* (pp. 120–143). Hershey, PA: IGI Global. doi:10.4018/978-1-5225-7479-8.ch007

Sharma, H. (2019). Skill Development, Skill Premium, and Technological Change: A Gender-Based Approach. In S. Bala & P. Singhal (Eds.), *Gender Issues in Technical and Vocational Education Programs* (pp. 200–217). Hershey, PA: IGI Global. doi:10.4018/978-1-5225-8443-8.ch010

Sibanda, F., & Humbe, B. P. (2020). "Prayers That Preyed": Critical Reflections on Robert Martin Gumbura's Church Saga in Zimbabwe. In B. Glimps & T. Ford (Eds.), *MeToo Issues in Religious-Based Institutions and Organizations* (pp. 144–169). Hershey, PA: IGI Global. doi:10.4018/978-1-5225-9195-5.ch006

Sibirskaya, A. (2019). Gender Issues and Priorities in Republic of Belarus. In S. Bala & P. Singhal (Eds.), *Gender Issues in Technical and Vocational Education Programs* (pp. 43–69). Hershey, PA: IGI Global. doi:10.4018/978-1-5225-8443-8.ch003

Singh, D. (2019). Gender Issues in Vocational Education and Training in India: Imperatives and Challenges. In S. Bala & P. Singhal (Eds.), *Gender Issues in Technical and Vocational Education Programs* (pp. 70–94). Hershey, PA: IGI Global. doi:10.4018/978-1-5225-8443-8.ch004

Sivakumar, M. (2020). Women in the Medical Field in India: A Developing Economy – A Dichotomous Picture. In M. Bellini & V. Papalois (Eds.), *Gender Equity in the Medical Profession* (pp. 230–236). Hershey, PA: IGI Global. doi:10.4018/978-1-5225-9599-1.ch016

Smith, T. T., & Johnson, N. N. (2019). Creating Support Systems for Black Women in Nontraditional STEM Career Paths. In U. Thomas & J. Drake (Eds.), *Women's Influence on Inclusion, Equity, and Diversity in STEM Fields* (pp. 108–142). Hershey, PA: IGI Global. doi:10.4018/978-1-5225-8870-2.ch005

Spencer, K. G. (2020). Men and Women Against the Other. In B. Glimps & T. Ford (Eds.), *MeToo Issues in Religious-Based Institutions and Organizations* (pp. 56–75). Hershey, PA: IGI Global. doi:10.4018/978-1-5225-9195-5.ch003

Statti, A. L., & Torres, K. M. (2019). Innovative Approaches to Traditional Mentoring Practices of Women in Higher Education. In H. Schnackenberg & D. Simard (Eds.), *Challenges and Opportunities for Women in Higher Education Leadership* (pp. 1–19). Hershey, PA: IGI Global. doi:10.4018/978-1-5225-7056-1.ch001

Storslee, J. (2019). Women Leaders Evolution in Our Society From the Perspective of a Son of a Women Leader. In H. Schnackenberg & D. Simard (Eds.), *Challenges and Opportunities for Women in Higher Education Leadership* (pp. 268–276). Hershey, PA: IGI Global. doi:10.4018/978-1-5225-7056-1.ch016

Sudan, F. K. (2019). Tapping Rural Women Entrepreneurship Through Self-Help Micro-Credit: Evidence and Lessons From Jammu and Kashmir, India. In F. Tomos, N. Kumar, N. Clifton, & D. Hyams-Ssekasi (Eds.), *Women Entrepreneurs and Strategic Decision Making in the Global Economy* (pp. 305–339). Hershey, PA: IGI Global. doi:10.4018/978-1-5225-7479-8.ch015

Tabassum, S. (2019). Gender Equity and Social Inclusion: Holistic Approach to Education, Training, and Labor Market Demand. In S. Bala & P. Singhal (Eds.), *Gender Issues in Technical and Vocational Education Programs* (pp. 119–137). Hershey, PA: IGI Global. doi:10.4018/978-1-5225-8443-8.ch006

Tan, B. U. (2020). The Effects of Values and Gender Role Perceptions on Attitudes Towards Women Managers. In B. Tan (Ed.), *Macro and Micro-Level Issues Surrounding Women in the Workforce: Emerging Research and Opportunities* (pp. 37–57). Hershey, PA: IGI Global. doi:10.4018/978-1-5225-9163-4.ch002

Tanrikulu, C. (2019). Understanding the Reasons for Gender Difference in Online Information Processing of Consumers: Based on Theories. In I. Williams, O. Millward, & R. Layton (Eds.), *Gender Gaps and the Social Inclusion Movement in ICT* (pp. 230–252). Hershey, PA: IGI Global. doi:10.4018/978-1-5225-7068-4.ch012

Tawiah, S., & Setlhodi, I. I. (2020). Introducing Information and Communication Technology Training for Rural Women in South Africa: Innovative Strategies for the Advancement of Livelihoods. *International Journal of Adult Education and Technology*, *11*(1), 45–59. doi:10.4018/IJAET.2020010103

Tepe, F. F. (2020). Researching Sensitive Topics: The Case of Sexism and Gender Discrimination at Turkish Universities. In B. Tan (Ed.), *Macro and Micro-Level Issues Surrounding Women in the Workforce: Emerging Research and Opportunities* (pp. 236–261). Hershey, PA: IGI Global. doi:10.4018/978-1-5225-9163-4.ch010

Tevis, T. (2019). The Challenges and Responsibilities of White Women in Leadership: Strategies for Interrogating Whiteness in Higher Education. In H. Schnackenberg & D. Simard (Eds.), *Challenges and Opportunities for Women in Higher Education Leadership* (pp. 125–143). Hershey, PA: IGI Global. doi:10.4018/978-1-5225-7056-1.ch008

Thomas, J. O., Minor, R., & Odemwingie, O. C. (2017). Exploring African American Middle-School Girls' Perceptions of Themselves as Game Designers. In Y. Rankin & J. Thomas (Eds.), *Moving Students of Color from Consumers to Producers of Technology* (pp. 49–61). Hershey, PA: IGI Global. doi:10.4018/978-1-5225-2005-4.ch003

Thomas, U., & Drake, J. (2019). Grace Under Fire. In U. Thomas & J. Drake (Eds.), *Women's Influence on Inclusion, Equity, and Diversity in STEM Fields* (pp. 169–193). Hershey, PA: IGI Global. doi:10.4018/978-1-5225-8870-2.ch007

Thomas, U. C., & Carter, K. W. (2019). Claws and All: Women of Color and the Pitfalls of Dominant Culture Leadership. In U. Thomas (Ed.), *Navigating Micro-Aggressions Toward Women in Higher Education* (pp. 122–143). Hershey, PA: IGI Global. doi:10.4018/978-1-5225-5942-9.ch006

Thompson, P., & Zang, W. (2019). Exports and Female Entrepreneurs: Foreign Influence and Domestic SME Exports. In F. Tomos, N. Kumar, N. Clifton, & D. Hyams-Ssekasi (Eds.), *Women Entrepreneurs and Strategic Decision Making in the Global Economy* (pp. 167–194). Hershey, PA: IGI Global. doi:10.4018/978-1-5225-7479-8.ch009

Tomos, F., Clifton, N., Ssekasi, D. H., & Balan, O. C. (2019). Inequalities: A Concern for Capitalism and Global Strategy. In F. Tomos, N. Kumar, N. Clifton, & D. Hyams-Ssekasi (Eds.), *Women Entrepreneurs and Strategic Decision Making in the Global Economy* (pp. 1–19). Hershey, PA: IGI Global. doi:10.4018/978-1-5225-7479-8.ch001

Tomos, F., Clifton, N., Thurairaj, S., & Balan, O. C. (2019). Perspectives on Women's Entrepreneurial Learning and the Evolution of Female Entrepreneurship Research. In F. Tomos, N. Kumar, N. Clifton, & D. Hyams-Ssekasi (Eds.), *Women Entrepreneurs and Strategic Decision Making in the Global Economy* (pp. 232–257). Hershey, PA: IGI Global. doi:10.4018/978-1-5225-7479-8.ch012

Tomos, F., & Thurairaj, S. (2019). New Emerging Technologies and ICT and Their Impact on Women Entrepreneurs' Success. In F. Tomos, N. Kumar, N. Clifton, & D. Hyams-Ssekasi (Eds.), *Women Entrepreneurs and Strategic Decision Making in the Global Economy* (pp. 144–166). Hershey, PA: IGI Global. doi:10.4018/978-1-5225-7479-8.ch008

Tunç, A. Ö., & Alkan, D. P. (2019). Women Entrepreneurship in Turkey as an Emerging Economy: Past, Present, and Future. In F. Tomos, N. Kumar, N. Clifton, & D. Hyams-Ssekasi (Eds.), *Women Entrepreneurs and Strategic Decision Making in the Global Economy* (pp. 40–62). Hershey, PA: IGI Global. doi:10.4018/978-1-5225-7479-8.ch003

Vig, S., & Dent, P. (2020). Training Martians and Venusians on Jupiter: Gender Equity in Medical Training. In M. Bellini & V. Papalois (Eds.), *Gender Equity in the Medical Profession* (pp. 51–63). Hershey, PA: IGI Global. doi:10.4018/978-1-5225-9599-1.ch004

Walls, T. E., Cornejo, M. N., Plachowski, T. J., Reid, E. K., & Park, S. (2018). Sowing Seeds of Justice: Feminists' Reflections on Teaching for Social Justice in the Southwest. In M. Grant (Ed.), *Equity, Equality, and Reform in Contemporary Public Education* (pp. 174–196). Hershey, PA: IGI Global. doi:10.4018/978-1-5225-4960-4.ch009

West, N. M., & Jones, T. B. (2019). Architects of Change in the Ivory Tower: Recasting the Role of Black Women Engaged in Higher Education Professional Counterspaces. In U. Thomas (Ed.), *Navigating Micro-Aggressions Toward Women in Higher Education* (pp. 23–52). Hershey, PA: IGI Global. doi:10.4018/978-1-5225-5942-9.ch002

Wheaton, M. M., & Kezar, A. (2019). Interlocking Systems of Oppression: Women Navigating Higher Education Leadership. In H. Schnackenberg & D. Simard (Eds.), *Challenges and Opportunities for Women in Higher Education Leadership* (pp. 61–83). Hershey, PA: IGI Global. doi:10.4018/978-1-5225-7056-1.ch005

Williams, I. (2019). Personality and ICT Adoption: The Call for More Research. In I. Williams, O. Millward, & R. Layton (Eds.), *Gender Gaps and the Social Inclusion Movement in ICT* (pp. 104–129). Hershey, PA: IGI Global. doi:10.4018/978-1-5225-7068-4.ch006

Wireko, J. K. (2019). Internet Addiction and Fear of Missing Out: The Gender Difference of Tertiary Students in Ghana. In I. Williams, O. Millward, & R. Layton (Eds.), *Gender Gaps and the Social Inclusion Movement in ICT* (pp. 253–268). Hershey, PA: IGI Global. doi:10.4018/978-1-5225-7068-4.ch013

Woodley, X. M. (2019). Voices in the Desert: Black Women Faculty in the American Southwest. *International Journal of Bias, Identity and Diversities in Education, 4*(2), 1–13. doi:10.4018/IJBIDE.2019070101

Yang, D., & Spear, C. E. (2017). The Effects of Faculty Status, Faculty Gender, Field of Study, and Class Size on the Use of Blogs, Wikis, and Discussion Boards. *International Journal of Information and Communication Technology Education, 13*(2), 52–64. doi:10.4018/IJICTE.2017040105

Yanıkkaya, B. (2020). What Does It Mean to Be a "Woman Leader" in Academia?: Imposing Patriarchal and Capitalist Ways of Leadership on Women. In B. Tan (Ed.), *Macro and Micro-Level Issues Surrounding Women in the Workforce: Emerging Research and Opportunities* (pp. 210–235). Hershey, PA: IGI Global. doi:10.4018/978-1-5225-9163-4.ch009

Yaokumah, W., Totimeh, F., & Kumah, P. (2019). Gender Differences in Preferences and Proclivities for ICT Tools and Online Services. In I. Williams, O. Millward, & R. Layton (Eds.), *Gender Gaps and the Social Inclusion Movement in ICT* (pp. 87–103). Hershey, PA: IGI Global. doi:10.4018/978-1-5225-7068-4.ch005

Yasim, Y. K. (2019). Relation Between Work-Family Life Conflict and Organizational Cynicism in Women: A Research in Banking Sector. In E. Pereira & P. Paoloni (Eds.), *Handbook of Research on Women in Management and the Global Labor Market* (pp. 113–135). Hershey, PA: IGI Global. doi:10.4018/978-1-5225-9171-9.ch006

Yıldırım, S., Yıldırım, D. Ç., & Çoltu, S. (2019). The Role of Education in Women's Career Life in Emerging Economies: A Case of Turkey. In E. Pereira & P. Paoloni (Eds.), *Handbook of Research on Women in Management and the Global Labor Market* (pp. 46–62). Hershey, PA: IGI Global. doi:10.4018/978-1-5225-9171-9.ch003

Yong, S. (2017). Digital Native Students: Gender Differences in Mathematics and Gaming. *International Journal of Virtual and Personal Learning Environments, 7*(2), 13–26. doi:10.4018/IJVPLE.2017070102

Zakria, M., Nova dos Santos, P. M., Moreira, A. C., & Mota, J. (2019). Exploring Female Entrepreneurship: Addressing Some Misconceptions. In E. Pereira & P. Paoloni (Eds.), *Handbook of Research on Women in Management and the Global Labor Market* (pp. 239–258). Hershey, PA: IGI Global. doi:10.4018/978-1-5225-9171-9.ch012

Zappala-Piemme, K. E. (2019). Women in Leadership: Summary for Success. In H. Schnackenberg & D. Simard (Eds.), *Challenges and Opportunities for Women in Higher Education Leadership* (pp. 277–297). Hershey, PA: IGI Global. doi:10.4018/978-1-5225-7056-1.ch017

Zhou, C. (2019). Meeting Gender Gaps in Information and Communication Technology (ICT): How Can Creativity Make a Difference? In I. Williams, O. Millward, & R. Layton (Eds.), *Gender Gaps and the Social Inclusion Movement in ICT* (pp. 212–229). Hershey, PA: IGI Global. doi:10.4018/978-1-5225-7068-4.ch011

Zin, S. M., & Manaf, K. A. (2019). Role of Intellectual Capital in Women Entrepreneurs' Business Performance. In F. Tomos, N. Kumar, N. Clifton, & D. Hyams-Ssekasi (Eds.), *Women Entrepreneurs and Strategic Decision Making in the Global Economy* (pp. 209–230). Hershey, PA: IGI Global. doi:10.4018/978-1-5225-7479-8.ch011

About the Contributors

Pamela M. Leggett-Robinson has 15+ years of higher education experience which includes STEM academic and student success/support programming, STEM Leadership Coaching, institutional strategic planning, data analytics, educational programming (start-ups), and program evaluation. Her diverse skill set is a result of having served as an academic administrator, project manager, principal investigator/ project director for STEM student and community initiatives, and lobbyist for K-20 science funding on Capitol Hill. She is a Certified Associate of Project Management and brings an exceptional level of enthusiasm, dedication, and nuanced perspective to each STEM program/project she serves. Dr. Leggett-Robinson holds a B.S. in Chemistry from Georgia State University, M.S. in Bio-Inorganic Chemistry from Tennessee Technological University, and a Ph.D. in Physical Organic Chemistry from Georgia State University. She is the Founder and Executive Director of PLR Consulting in Atlanta, GA. PLR Consulting focuses on STEM program development, project management, and evaluation. Dr. Leggett-Robinson's primary research and scientific presentations focus on student support programs in STEM undergraduate education and support programming for women of color in STEM professions. She has received multiple awards for her STEM service to students and garnered funds in excess of $3 million dollars from both NIH and NSF for broadening participation in STEM Undergraduate Education. Her distinguished record of STEM programmatic success is well documented in publications and presentations. Through speaking engagements, Dr. Leggett-Robinson works with national, regional, and local organizations to create and implement safe STEM ecosystems for women of color.

Brandi Campbell Villa would say that she has been a researcher since childhood, gradually honing her skills through formal and informal education. By the time she finished her undergraduate studies, she had also become an educator. Brandi did her graduate research on two fronts – in the field of applied and environmental microbiology as well as in program evaluation. With more than a decade of teaching experience, Brandi brings an educator and researcher's perspective to the design and implementation of our projects. Her evaluation and non-profit work has taken her around the globe to contribute to national and international projects, including bilingual programs in English and Spanish. She excels at working with people from different backgrounds, cultures, and even different languages. She has taught scientific writing for years, and has collaborated with many of our clients to write and edit successful grant proposals, conference presentations, and journal publications. In addition to her passion for all aspects of STEM education and health, Brandi particularly enjoys challenges related to evaluation design, scientific communication for the lay audience, reporting and data visualization, and meeting facilitation.

* * *

Jessica T. DeCuir-Gunby is a Professor of Educational Psychology and University Faculty Scholar in the Department of Teacher Education and Learning Sciences at NC State University. She also serves as the department's Director of Graduate Programs and Interim Associate Department Head. In addition, she currently serves on the editorial boards for Contemporary Educational Psychology, Educational Psychologist, and the American Educational Research Journal, where she formerly served as an associate editor. DeCuir-Gunby's research interests include race and racial identity development, critical race theory, mixed methods research, and emotions in education. Her work has been featured in top-tier journals such as Contemporary Educational Psychology, Educational Psychologist, Educational Researcher, and Review of Educational Research, among others. She has also co-authored/co-edited four books.

Christine Grant, a full Professor of Chemical and Biomolecular Engineering, has served as the inaugural Associate Dean of Faculty Advancement in NC State's College of Engineering for 11 years. She obtained her degrees in Chemical Engineering (B.S., Brown University; M.S. and Ph.D., Georgia Institute of Technology) prior to joining NC State over thirty years ago. Grant has provided professional leadership in the American Institute of Chemical Engineers (AIChE) as: member of the Board of Directors and the Chemical Technology Operating Council, Chair of the Minority Affairs Committee, member of Chemical Technology Operating Council, National Awards Committee and an AIChE Fellow. Grant is one of less than ten African-American women CHE full professors in the U.S. at the rank of full professor in chemical engineering. A recipient of the AAAS Mentor Award and the NSF Presidential Award for Excellence in Science, Math and Engineering Mentoring (PAESMEM); Grant's been recognized for broadening the participation, promotion and retention of underrepresented minorities (URM) and women in STEM with: AIChE Minority Affairs Committee Pioneers of Diversity Award, Council for Chemical Research Diversity Award, the American Chemical Society's

SERMACS Diversity Award, and NOBCChE's Lloyd M. Fergusson Young Scientist Awards. A Life Member of the Society of Women Engineers, NSBE and AIChE, she's been: a Visiting Senior Scholar at AAAS, an Expert in the NSF Engineering Directorate, and a National Academy of Engineering Boeing Senior Fellow. Grant conducts STEM Resilience coaching and professional development workshops; her book, "Success Strategies from Women in STEM: A Portable Mentor" is the culmination of Grant's 30+ years work in STEM.

Tyresa Jackson is a graduate student at Teachers College, Columbia University. Prior to matriculating into Teachers College she earned a Masters of Public Administration from John Jay College of Criminal Justice. Her professional experiences include being a graduate assistant at John Jay College of Criminal Justice for Women in the Public Sector, and being a research assistant on a project that assessed disaster risk management and resilience in Kenya, Haiti, and Jamaica. In her free time, she is a special events volunteer for Girls on the Run – New York City Chapter and she serves as the secretary for the American Society for Public Administration (ASPA) New York Metro Chapter. Her interests include closing the education-to-prison pipeline, education reform, and increasing the number of women in STEM (Science, Technology, Engineering, and Math).

Sherine Obare is the Dean of the Joint School of Nanoscience and Nanoengineering, an innovative collaboration between the University of North Carolina at Greensboro (UNCG) and North Carolina A&T State University. She is also a Professor of Nanoscience at the University of North Carolina at Greensboro. She received a B.S. in Chemistry from West Virginia State University, obtained a Ph.D. in Chemistry from the University of South Carolina, and thereafter, was a Camille and Henry Dreyfus Postdoctoral fellow at The Johns Hopkins University. Obare is an environmental chemist whose research focuses on nanotechnology. Her work has led to innovations in the detection and remediation of environmental contaminants, designing nanoscale materials for drug delivery, improved healthcare, and alternative energy, as well as understanding the fate, transport and toxicity of anthropogenic nanomaterials. She began her career at Western Michigan University as Assistant Professor of Chemistry in 2004, was promoted to Associate Professor in 2009, and to Full Professor in 2014. At Western Michigan University, she served in leadership roles as Associate Dean for Research in the College of Arts and Sciences from 2015 – 2016 and as Associate Vice President for research from 2016 - 2018. Dr. Obare's research program has been funded by the National Science Foundation, the Department of Defense, the Army Research Office, the National Institutes of Health, the Department of Education, and the Michigan Economic Development Corporation. Dr. Obare is a 2019 Fellow of the American Chemical Society, a 2018 Fellow of the

Council of Research of the Association of Public and Land-Grant Universities, the recipient of the 2009 International Union of Pure and Applied Chemistry (IUPAC) Young Observer Award, the National Science Foundation CAREER award, and the 2010 NSF Division of Materials Research American Competitiveness and Innovation (ACI) Fellowship. In 2013, Obare was named as one of the top 25 Women Professors in Michigan. Her research work has been published widely in scientific journals and in books. She has mentored a large number of graduate, undergraduate and high school students in research. She has been highly active promoting science to elementary and middle school students in the locally and nationally. She is the co-Editor of 'Green Technologies for the Environment'. Dr. Obare serves as an Associate Editor for the Journal of Nanomaterials.

Cailisha L. Petty, PhD, is the Master of Arts in Teaching Program Coordinator at North Carolina A&T State University. She earned a PhD in Curriculum and Teaching from the University of North Carolina at Greensboro. Her research interests include equity in Science Education for Students of Color and preparing engaging, highly qualified STEM Educators.

Nichole L. Powell is an Associate Professor of Chemistry at Oxford College of Emory University. She is passionate about exposing students to the research process by challenging them with problems that are difficult, yet within their reach, and allowing them to learn by making their own mistakes. Her research interests include the synthesis and evaluation of natural product analogs as potential therapeutic agents as well as the incorporation of scientific inquiry into the first and second-year curriculum and the investigation of ways for improving the teaching and learning of chemistry. An advocate for science literacy and diversity initiatives, she serves as the Chair of the Minority Affairs Committee - Georgia Section of the American Chemical Society and is also involved in efforts at Emory University related to Diversity, Equity and Inclusion. She has received recognition for excellence in teaching and exceptional contribution to the education of Oxford College students.

Pamela E. Scott-Johnson, Ph.D., is dean of the College of Natural and Social Sciences at California State University Los Angeles. Prior to that, she served as the interim dean of the College of Liberal Arts and former chairperson of the Department of Psychology at Morgan State University. She earned a B. A. in psychology and graduated magna cum laude from Spelman College, and an M.A. and Ph.D. in psychology and neuroscience from Princeton University. Her research interests include chemosensory sciences (bitter taste; olfactory epithelial response), eating disorders, body image and sensory systems. Her work has also facilitated preparing more diverse and inclusive access for those in the field of psychology and neuroscience. Her administrative goals are ensuring quality education and access for all students; identifying and developing leadership talent among faculty; creating a diverse and inclusive environment to meet the institutional mission and priorities; and increasing institutional capacity.

Angela Michelle White has served as an educator for 15 years at various levels and currently serves as the Assistant Dean of Student Success for the College of Science and Technology at North Carolina Agricultural and Technical State University. Her current research interests, publications, and presentations give attention to racial identity, science identity, science self-efficacy, metacognition, critical race theory, mixed methods research and STEM achievement of African American students. Her work has been featured in Contemporary Educational Psychology and Race, Ethnicity and Education.

Index

Ensure Quality Research is Introduced to the Academic Community

Become an IGI Global Reviewer for Authored Book Projects

The overall success of an authored book project is dependent on quality and timely reviews.

In this competitive age of scholarly publishing, constructive and timely feedback significantly expedites the turnaround time of manuscripts from submission to acceptance, allowing the publication and discovery of forward-thinking research at a much more expeditious rate. Several IGI Global authored book projects are currently seeking highly-qualified experts in the field to fill vacancies on their respective editorial review boards:

Applications and Inquiries may be sent to:
development@igi-global.com

Applicants must have a doctorate (or an equivalent degree) as well as publishing and reviewing experience. Reviewers are asked to complete the open-ended evaluation questions with as much detail as possible in a timely, collegial, and constructive manner. All reviewers' tenures run for one-year terms on the editorial review boards and are expected to complete at least three reviews per term. Upon successful completion of this term, reviewers can be considered for an additional term.

If you have a colleague that may be interested in this opportunity, we encourage you to share this information with them.

www.ingramcontent.com/pod-product-compliance
Lightning Source LLC
Chambersburg PA
CBHW082107111025
33823CB00004B/25